FACT AND THEORY

W. M. O'NEIL

Fact and Theory

An aspect of the philosophy of science

SYDNEY UNIVERSITY PRESS

SYDNEY UNIVERSITY PRESS
Press Building, University of Sydney

NEW ZEALAND   Price Milburn and Company Limited
GREAT BRITAIN   Methuen and Company Limited, London
and their agents overseas

First published 1969
Copyright 1969 by W. M. O'Neil

National Library of Australia registry number AUS 68-2391
Library of Congress Catalog Card Number 68-27809
SBN 424 05800 6

This book is supported by money from
THE ELEANOR SOPHIA WOOD BEQUEST

Printed in Australia at The Griffin Press, Adelaide
and registered in Australia for transmission by post as a book

Contents

List of Figures

vii

List of Tables

Preface

In the Spring Semester of 1960, I had the privilege as Visiting Professor under the Tallman Foundation of giving a course in the philosophy of science at Bowdoin College, Maine. I set out to interest students majoring in the several sciences, adopting the pedagogic device of raising philosophical issues through a consideration of some topics in the history of science. The device seemed to work though at the time I felt that I was clumsy in its use. Later I tried it again, with some changes in the material, in a shorter course given to senior undergraduates in the Faculty of Science in the University of Sydney. The response on each occasion not only from the students but also from others who had dropped in to listen, if not to mock, has led me to work over the material once again in order to produce a small book that may introduce some of the problems in the philosophy of science to several types of reader.

I have continued to keep the senior science undergraduate foremost in mind. There are many good books written by professional philosophers for them to read. However, these books concentrate on problems which the philosophers see to be of main importance but which at first sight the science student is likely to regard as having too little bearing on science as he has encountered it. I hope that my use of cases drawn from the history of science may convince him that there is some relevance in the philosophers' problems. As the case studies used are in the main old and hence are now easy to understand, the intelligent layman without up-to-date scientific knowledge may also find some enlightenment in the material presented. Finally, I am bold enough to think that some students of the philosophy of science may be served in a peripheral way. All too often they lack sufficient knowledge of the sciences and find themselves trying to follow their teachers through the intricacies of formal issues arising in the more sophisticated levels of modern science. It is hoped that this discussion may help them in that it starts at the grass roots. It will almost certainly infuriate them as it never gets far above the ground and often draws back from the giddier heights of philosophical discussion.

I have conceived the role of this book to be not that of the preacher or the teacher, whether to the converted or to the pagan, but instead that of the man with the megaphone at the front of the tent on the fairground. It is not here to put on the show but to induce some passers-by to make the decision to enter and to see the show itself. No doubt the talker-at-the-front finds some satisfaction in giving a little entertainment and instruction to those who pause a while but do not enter. I shall not be fully satisfied, however, unless those who read this book go on to think and to read more about the problems it raises and to which it presents no more than partial and tentative answers. This is perhaps an immodest aim for so modest a book.

In addition to my great indebtedness to the authors of the many books from which I have derived the material which follows, I have many debts of a more personal kind. Professor M. G. Taylor, Dr Harley Wood, Professor Ian Ross and Professor R. J. Walsh read part or all of the manuscript and saved me from putting into print some of my grosser errors in and misunderstandings of physiology, astronomy, chemistry and genetics respectively; an unidentified publisher's reader put me on to Fleming's correction of the modern misconception of Galen's views about the motion of the blood. Mr D. C. Stove, who read the whole with a philosopher's eye, drew my attention to many gaffes I have since removed. None of them should be held responsible for the errors that remain, if for no other reason than that, being obstinate in my heresies, I have in several places refused to accept their advice. Once again my wife has with patience persuaded me to remove some of the worst of my habitual errors of punctuation and expression, and has with equal persistence encouraged me to bring to some sort of conclusion a task on which I had been engaged for too many years.

W. M. O'NEIL

UNIVERSITY OF SYDNEY
April 1968

INTRODUCTION

A little acquaintance with science in any of its branches reveals that it is in part factual and in part theoretical. The facts in a science consist of what the scientist observes and the theories of what he supposes or invents. His suppositions are made sometimes to fill in gaps in his factual knowledge, sometimes to increase his understanding of his facts and sometimes for other reasons. The account of the many classes of living organisms, with their complex sets of similarities and differences, and the account of the fossil record in the rocks of a succession of long extinct species are largely statements of what anyone with the interest, time and the needed skill can observe for himself. The theory of evolution, which asserts that the several species originated through a long drawn out process involving natural variation, natural selection and inheritance, is not an account of a process which anyone has observed from beginning to end. It is a supposition, though not a complete supposition for it incorporates some observed facts. It is a supposition, nevertheless, as a whole. We have come to believe it to be true, to be as much a fact, in another sense of that word, as the descriptive similarities and differences between the species. But unlike that descriptive factual material, it has been invented and not observed. The aim of its inventors was to bring some sense, some intelligibility into what would otherwise have been brute, senseless facts. Whereas observation provides *an account of* the species and their relationships, the evolutionary theory of Darwin and others *accounts for* these facts.

This much is easy enough to recognize. There remain, however, many philosophic questions about the distinction between fact and theory and about the roles of fact and of theory in science. For example, do facts and theories differ merely in the way we arrive at them, that is, the former by observing and the latter by supposing? Or are they quite different kinds of knowledge? Facts could be matters that are true in nature, that is situations which really exist, whereas it could be that theories may not give us information about the world but instead provide us with a conceptual framework for understanding what we observe. Or are they different in their reliability or trustworthiness? Sometimes scientists treat facts as the bedrock upon which they must build, abandoning a theory whenever the facts do not support it. Sometimes, however, we find scientists doubting a stray fact because it does not conform to a well-established theory. It could be, of course, that there are several sorts of facts and several sorts of theories, and that one has to answer the foregoing questions differently for each sort. When we attempt to answer such questions as these, others emerge. Probing the wound aggravates the hurt, whatever ameliorative effects may be produced in the long run.

Some scientists see little point in bothering with these questions at all. They think that anyone wanting to engage in science should simply roll up his sleeves and get on with his task—that it is worse than useless for him to stand back in contemplation of science in process. They may well be right. Perhaps the centipede will stumble, or even be immobilized, should it pause to ask itself which foot should be lifted and which put down next. On the other hand, there may be little harm and possibly some profit for a centipede, which does not at the moment wish to walk, in asking these questions of one of its fellows which is in locomotion.

Sometimes the philosophy of science is cultivated by scientists. More often it is cultivated by professional philosophers. Because philosophy has at times seemed to be concerned with what ought to be rather than with what is, some of those who study science lifting and putting down its feet prefer to name their activity 'the science of science'. However, the philosophy (or science) of science is not quite the study of what science actually does and it is not quite the laying down of what science ought to do. The professional philosopher, it must be admitted, is prone to raise questions and to make statements which seem to the scientist to have tangential, if any, reference to science. Consequently, the consideration of one area of the philosophy of science, which follows, will be made through a series of case studies drawn from the history of the sciences. These illustrative cases, which constitute one strand of this book, have been chosen because they draw attention to various issues about facts and theories and about the relation between them. The other intertwined strand of comment attempts to draw together, to add to and to interpret the general points brought out in the case studies. The more usual procedure in discussions of these topics is to begin and to end with generalities illustrated from time to time with specific examples drawn from science. The procedure adopted here seems to be a little closer to the procedure characteristic of science itself.

Part One

THE MOTION OF THE BLOOD

1 *CASE STUDY*

Our first and quite brief historical case study in the philosophy of science begins with Galen (*circa* AD 130-200) and ends shortly after Harvey (AD 1578-1657). Harvey is credited with the discovery of the circulation of the blood. He certainly asserted that the blood circulated and provided evidence in support of his view; in doing so he supplemented and corrected, if not displaced, the views of Galen on the motion of the blood which had been accepted as authoritative during the fourteen centuries which separated them.

Galen was born in Pergamon, where at about fourteen years of age he began the study of philosophy. Later he turned to medical studies and in the pursuit of further knowledge in that field he travelled to Smyrna, Corinth and Alexandria. In his early thirties he went to Rome, where he practised medicine for many years. He became physician to the Emperor Marcus Aurelius and gave public lectures in Rome especially on anatomy. He was a careful, systematic investigator and wrote at length on anatomical, physiological and general medical topics. He did a great deal of dissection in order to further his anatomical knowledge, using in particular the Barbary ape and the rhesus monkey for this purpose. He had a good deal to say about the heart and the elongated blood vessels branching from it and about the origin, flow and function of the blood that was contained in these vessels.

A simplified and reduced version of Galen's teachings will serve our present purpose. Some of what he had to say was almost entirely a report of what he observed, some was largely what he supposed in order to understand what he observed, and in between these two extremes were various degrees of admixture of observation and supposition. In order to emphasize these differences, I shall use the words 'he observed', 'he supposed' and 'it seemed to him' to denote the ends and the middle of the spectrum of sources of his knowledge. At this remote time it is easier to set down what he had to say about the structure of the vascular system than it is to set down what he

5

had to say about the motion of the blood. Possibly he was clearer about the former.

He observed the structure of the walls of the heart, but it seemed to him that while it had something in common with the skeletal muscles it had important differences. He observed that the heart had two chambers and he supposed, on the basis of the species he had examined, that this was true of all air breathing animals. He observed that the upper part (the little ear or auricle) of each chamber was separated from the main chamber (ventricle) by a membranous structure we now call the valve, and he supposed that it acted so as to produce a one-way flow of the blood at that point. He observed the elongated vessels which led off with ever increasing branching, from the two chambers of the heart. Considering for the moment those vessels linking the heart with parts of the body other than the lungs, Galen observed that those linked with the left chamber of the heart pulsed in concert with the heart, whereas those linked with the right chamber did not. He called the former arteries and the latter veins. Earlier observers had reported that whereas the veins contained blood, the arteries did not. Galen produced decisive observations showing that the arteries contained blood. He observed too that arterial blood was brighter in colour than venous blood and he supposed that it had been aerated (converting the 'natural spirits' of venous blood into 'vital spirits') in some way by the incorporation of air taken into the lungs. He was puzzled in his attempt to classify as veins and arteries the blood vessels linking heart and lungs, and took refuge in such compromises as labelling the vessels, which we now call the pulmonary veins (leading from lungs to left ventricle), the venous arteries. Had he been able to detect a pulse in those vessels, he would have had no doubt that they were arteries; because they joined the heart on its left side, it seemed to him that they should be arteries. However, he could detect no pulse in them and they had the thin walls typical of veins. Out of similar puzzlement he called the vessel, which we now call the pulmonary artery, the arterial vein.

It seemed to him (perhaps I should say, he observed) that the blood moved in the heart and other blood vessels. It also seemed to him (perhaps I should say, he supposed) that the blood brought nutriment from the intestines to the various organs of the body and carried away from them noxious vapours which were breathed out from the lungs. In addition to this latter task of a sewage kind, he supposed that the

6

function of the blood was that of irrigation, bringing nutriment, 'natural spirits' and 'vital spirits', to the various body organs. Because the liver stood so prominently between the intestines, from which it seemed nutriment was derived, and the heart, Galen supposed that blood was made by it and that it was the centre of the venous system. Though modern historians of medicine, including the redoubtable Singer, have asserted that Galen supposed that the blood ebbed and flowed in the venous system, Fleming has recently argued in a most convincing way that he did not, indeed that he contended that the valve at the entrance of the right ventricle and the valve at the base of the 'arterial vein' prevented any but a minimal backward flow of the blood from its main direction namely vena cava to right ventricle and thence by the 'arterial vein' to the lungs. Now if the blood were made in the liver and distributed from there into the venous system, how did it get into the arterial system? Historians of medicine have always stressed Galen's assumption of invisible pores in the septum of the heart as his answer to this question. Fleming contends that he provided at least one other route and possibly two. First, Galen supposed that there were 'anastomoses' or tiny openings between the veins and the arteries throughout the body allowing an exchange of blood between the two systems. Second, he supposed that through such 'anastomoses' in the lungs, blood flowing from the right ventricle into the 'arterial vein' passed into the 'venous arteries'. He seems not to have asserted that it flowed thence into the left ventricle. Indeed he seems to have been silent on what happened to blood entering the 'venous arteries' through 'anastomoses' in the lungs. He did, however, suppose that air inspired into the lungs flowed through the 'venous arteries' into the left ventricle where it contributed 'vital spirits' to the blood. He further supposed that the waste products were given up by the blood in the left ventricle and passed 'upstream' along the 'venous arteries' to the lungs whence they were expired in the breath.

This somewhat vague account of the motion of the blood is both different in detail from the theory propounded by Harvey and falls far short of the notion of circulation. Though Harvey makes no suggestion that his theory was contrary to Galen's incomplete account, some doubts about Galen's assumptions and some discoveries in the century before Harvey were probably important for him.

Vesalius in the mid-sixteenth century questioned Galen's supposition that blood seeped through pores in the septum of the heart. The material of the septum appeared to him to be no different from that of the outer walls of the heart which did not allow any seepage of blood. Vesalius was, as it were, asking: 'Why suppose pores in the tissue when you can not see them, and when that tissue appears no different from tissue in which you do not suppose pores?' This is a question of a type always worth asking though we shall see that it does not do to answer it too hastily. At about the same time as Vesalius was throwing doubt on a seepage of blood through pores in the septum of the heart, Servetus vigorously asserted the importance of the pathway through the lungs. In particular he maintained that the blood was endowed with 'vital spirits' from the air and gave up its waste products before it completed, if not before it began, its journey along the 'venous arteries' to the left ventricle. He is often credited with the discovery of the lesser or pulmonary circulation. But there were others who anticipated him, although none of them may have stated the matter as clearly as he did. What he did state was not strictly a 'circulation' as it was merely an arc of the circle Harvey first made clear. In any case, it may be doubted that Servetus or his predecessors *discovered* the so-called pulmonary circulation. Those who supposed the Great South Land, *Terra Australis*, were not the discoverers of Australia. That title belongs to a number of Dutchmen, who overshot their mark before turning north to the East Indies, and to Dampier, Cook and others who sailed along its shores.

Another discovery by Fabricius, reported in 1574, was important for Harvey. Fabricius showed that the membranes distributed at intervals on the inside of the veins operated so as to allow the blood to flow freely in one direction but to hinder (he thought only partially) its flow in the other. He supposed that the valves prevented the blood falling in its entirety into the lower extremities of the body.

Though independent and careful observations were being made again after some fourteen centuries during which Galen's words were almost law, what Harvey inherited was a complex of observation and supposition in which one blended into the other without clear distinction. All this tradition, whether factual or theoretical, was almost as much a burden as a useful and usable possession. Harvey's revolution of thought about the flow of the blood depended upon his careful and intelligent observation which made the facts clearer to him, upon his critical analysis of the implications of earlier interpretations and upon his comparison of these implications with the facts as he knew them.

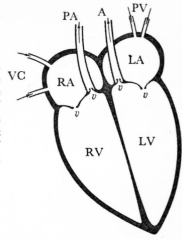

Fig. 1.1 Diagram of the mammalian heart. The vessels marked VC are the superior and inferior vena cava, PA the pulmonary artery, A the aorta, and PV the pulmonary veins. The chambers marked RA and LA are the right and left auricles respectively, and RV and LV the right and left ventricles respectively. The valves are marked *v*.

His main observations were made on living animals whose thoracic cavities had been opened. He found the cold-blooded animals especially useful as their slow heart beat gave him a better opportunity to see what was happening. By holding the heart as it beat, he observed that it hardened as it contracted. By analogy with the skeletal muscles, it seemed to him that the heart was a hollow muscle. He observed that the contraction of the heart was synchronous with the distension of the arteries. He observed too that when an artery was punctured blood spurted intermittently from the puncture and synchronously with the heart contractions. He supposed, therefore, that the heart was a hollow muscular pump. He observed that the contraction of the heart began at the auricles. He supposed, therefore, that the contraction of the auricles forced blood into the ventricles, and that the contraction of the ventricles forced it into the arteries. The views he inherited together with his own observations left him in no doubt that the thin membranes in the veins, and the analogous structures separating auricle and ventricle and at the roots of the arteries stemming from the heart, were valves allowing the blood to flow one way and absolutely preventing its flow in the opposite direction. The direction of flow was from the smaller to the larger veins, to the right auricle, to right ventricle then by the pulmonary artery through the lungs and through the pulmonary vein to the left auricle, to the left ventricle and finally through the aorta into the smaller arteries. Harvey did a simple calculation about

9

the amount of blood passing through the heart in a unit of time. He estimated that the human heart held about two ounces of blood. Assuming that it emptied itself at each beat, then at a beat rate of 72 per minute, the amount of blood passing through the heart was $2 \times 72 = 144$ ounces per minute or $144 \times 60 = 8,640$ ounces or 540 pounds per hour. He wondered where all this blood could come from and where it could go to in so short a time. Unless the blood circulated around the whole cardio-vascular system, blood weighing about three times the weight of an average man would have to be produced in an hour by the liver and absorbed in the same period by the tissues it irrigated. This was implausible to Harvey. He could think of no alternative to the supposition that the fine branched ends of the arteries were linked to the fine branched ends of the veins by, to him invisible, pores in the flesh. The heart, according to his view, pumped blood into the arteries, forcing it through the pores in the flesh into the veins from which it flowed back into the heart. He considered this view to be confirmed by the facts to be obtained by severing an artery of an animal. Blood could be seen to flow out in pulsations and not at a steady rate from the severed artery; as time progressed the flow could be seen to diminish, as though the supply of blood were running out; finally the animal would die apparently from lack of blood.

The translation of the final summarizing chapter of his book, *Anatomical dissertations concerning the motion of the heart and blood in animals*, is worth quoting:

And now I may be allowed to give in brief my view of the circulation of the blood, and to propose it for general adoption.

Since all things, both argument and ocular demonstration, show that the blood passes through the lungs and heart by the action of the [auricles and] ventricles, and is sent for distribution to all parts of the body, where it makes its way into the veins and pores of the flesh, and then flows by the veins from the circumference on every side to the centre, from the lesser to the greater veins, and is by them finally discharged into the vena cava and right auricle of the heart, and this in such a quantity or in such a flux and reflux thither by the arteries, hither by the veins, as cannot possibly be supplied by the ingesta, and is much greater than can be required for mere purposes of nutrition; it is absolutely necessary to conclude that the blood in the animal body is impelled in a circle, and is in a state of ceaseless motion; that this is the act or function which the heart performs by means of its pulse; and that it is the sole and only end of the motion and contraction of the heart.

Some thirty years after Harvey published his supposition of the pores in the flesh linking the arteries and the veins, Malpighi using an early microscope saw them. He saw first that the blood from the arteries in the lung of the frog separated out into tiny streams before flowing together into the veins. Later in a dried lung he found that these streams were guided by tiny tubules linking the arteries and the veins. Subsequently he observed capillaries in other parts of the body. Later microscopists with improved instruments were able to see the red corpuscles of the blood filing along the capillaries from the ends of the arteries to the veins. So much of what Harvey supposed, in order to fill the gaps left in his observed data and to make sense of the data he had, has since been observed that it may be said that his theory has been transformed into the observed fact of the circulation of the blood. His theory then was an anticipation of observations yet to be made. Not all theories are of this type as we shall discover in later case studies.

2 COMMENTS

SOME FEATURES OF SIMPLE THEORIES

Harvey's theory of the motion of the blood is a very simple type of theory. It is the supposition of matters which have not yet been observed but which might have been observed had circumstances been favourable. If Harvey had possessed a microscope as Malpighi did and had he thought to study with it the passage of the blood through the lungs of a frog as Malpighi did, he would have been able to observe the fine capillaries joining the end branches of the arteries and of the veins. But, as it happened, he was not favoured by such circumstances and so his only resort was to suppose that there were such linking pores in the flesh. He did this not simply because he was dissatisfied with the facts he had, and not at all because he was too impatient to wait for other facts to come to him or too lazy to look for them. He was puzzled by the incomplete set of facts he had. He thought he could detect the general outline of the picture of which the known facts constituted some of the parts. He had some of the pieces of the jigsaw in what seemed their right places but he had no more pieces with which to fill the remaining gaps. He proceeded then to imagine what the suspected missing pieces were like so that he could present a whole picture which made sense. The facts he had inherited together with those he had found for himself puzzled him, so he assumed additional items of 'information' in order to remove that puzzlement. For instance, granted that the heart worked as a pump, where did the enormous quantity of blood passing through the heart in an hour come from and where did it go to? This problem would be solved should it be the case that the arteries were linked by unseen pores in the flesh to the veins and that the blood went round and round over and over again.

Galen, of course, was likewise puzzled by the way blood, seemingly made in the liver which was linked only with the venous system of blood vessels, made its way into the arterial system where he had found it to be. This problem would be solved should it be the case that there were pores in the septum of the heart and 'anastomoses' between the veins and the arteries in the lungs and elsewhere in the body. Unfortunately for Galen, his supposition of the former has never received any observational support, whereas Harvey's supposition has been transformed into a fact of observation. It could have been otherwise. There is no reliable recipe for making suppositions which are later corroborated by observation. Suppositions such as those of Galen and of Harvey are made in order to complete the picture in an intelligible way. Later observation sometimes supplies the missing pieces and sometimes it does not. There is no way of telling in advance which way the story will end.

There is a difference, however, between an assumption of unseen pores, whether in the septum of the heart or in the flesh, as links between arteries and veins and an assumption of literally invisible pores. Had Harvey assumed pores in the flesh which were literally invisible, Malpighi's observations would have shown him to be wrong. If Galen had assumed that there were literally invisible pores in the septum of the heart, the failure to observe them would not have shown him to be wrong. No one has seen septal pores, but that does not show that there are not invisible septal pores. A theorist who means to prevent the overthrow of his theory may give it an inbuilt defence against unfavourable observations by endowing it with some unobservable crucial feature. This statement needs qualification for those theories which are not anticipations of what may, given favourable circumstances, be observed. Such theories will be encountered in later case studies.

When a theory of the type we have been discussing is propounded the evidence already available may support it to a greater or less extent. Harvey's theory of the circulation of the blood had much stronger supporting evidence at the time of its statement than did Servetus' theory of the pulmonary circulation. Harvey had more information about the blood vessels and the motion of blood in them than did Galen or Servetus. His theory was consistent with this information, and it answered puzzling questions prompted by this information in a way that no other theory, which had been advanced, did. The evidence did not prove his theory in the way that a theorem in geometry is proved by deduction from Euclid's definitions and axioms. In a geometrical proof, as in any form of strict logical inference, it is shown that acceptance of certain premises, or starting points, commits one to the acceptance of certain conclusions. Acceptance of Harvey's evidence, however, does not necessarily commit one to the acceptance of his theory—he exaggerated his case when he claimed just this. Had his evidence been inconsistent with the theory, then acceptance of the evidence would commit one to the rejection of his theory. The evidence was consistent with the theory without logically entailing it. That the theory also resolved problems posed by the evidence made the theory probable. That there was no alternative theory in like position rendered Harvey's even more probable.

Theories and facts are not completely and absolutely distinct. If they were, theories could not become facts as Harvey's theory has. What pass as facts are not pure observations. This is evident when we consider the way in which the products of observation are ordered and classified. Galen recognized that the pulmonary artery, as we now call it, had thick walls like the aorta and the arterial system branching from it, whereas the pulmonary veins, as we now call them, had thin walls like the veins. For Galen, however, the distinctive features about the arteries were the bright red (arterial) blood they contained, their pulsation and their being linked to the left chamber of the heart, and the distinctive features of the veins were the dark red (venous) blood they contained, their lack of pulsation and their being linked to the right chamber of the heart. Following Harvey, we classify the vessels as veins if they bring blood to the heart, whether it be dark or bright red in colour, and as arteries if they take blood from the heart. On sound functional grounds the arteries have thicker walls than the veins. The difference between the two classifications is not so much a matter of truth as of utility. If one emphasizes the structure of the walls of the pulmonary vessels, one classifies them in one way; if one empha- sizes the colour of the blood in them and the side of the heart to which they are linked, one classifies them in the other. This sort of thing has led some philosophers of science to talk about such parts of science as being arbitrary. However, it must be protested that the scientist is not free to classify things in any way that he wills. His classification, to be useful, must conform to the facts. It may, nevertheless, be one of a number of alternative ways of classifying complex factual data—the one that is most valuable and most illuminating. The question remaining is whether the classification is in some other sense true as well as conforming to the facts and being useful.

Part Two

THE MOTIONS AND
SPACING OF THE PLANETS

3 *CASE STUDY*

The second case study will be much longer than the first both in the telling and in drawing lessons from it for the philosophy of science. It could begin with the theory of Ptolemy, who was an Alexandrian contemporary of Galen, and end with the alternative theory of Copernicus, a northern European contemporary of Vesalius. However, as the basic facts were established by the Babylonians long before Ptolemy began to theorize about those facts, and as Copernicus opened but did not close the modern chapter on planetary theory, it will be desirable to spread the story from before the beginning of the second millenium BC to the early nineteenth century AD. Apart from some mythological adornments which we shall largely ignore, the Babylonian contribution is predominantly factual. They contributed certain conventions and methods of measuring the phenomena and of summarizing the facts but the early theory is almost entirely Greek and Hellenistic. Most of our terminology even for matters of fact comes from this period of Greek theorizing and the later Arab preservation and supplementation of it.

The apparent motions of the stars

Let us begin then, many centuries before Christ, by the waters of Babylon with our eyes not downcast in weeping but raised in wonder to watch the procession of the stars across the heavens above. Perhaps the most basic fact to be observed would be the difference between the Sun-dominated day-time sky and the star-sprinkled night sky, with the Moon, in its fickle way, sometimes a pale wraith in the one and a radiant form in the other. Next perhaps we would notice that, with the exception of some stars in the northern Babylonian sky, the celestial bodies, that is, the Sun, the Moon, and the stars, rise towards the east, pass in an arc of a circle across the sky and set towards the west, some well to the north, some well to the south and the rest somewhere in between. The exceptions in the northern sky can be seen never to dip below the horizon, each tracing during the night an arc of a circle completely above the horizon and centred on a point about 32.5 degrees above the horizon due north. This point we may call the north celestial pole. These several observed facts about the diurnal motion of the celestial bodies are illustrated in Fig. 3.1. They leave us with little

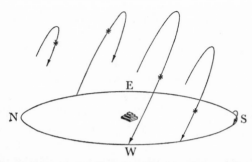

Fig. 3.1   Apparent celestial paths seen from Babylon. Notice the circumpolar path to the north, and the southward tilt of the paths. From a site north of Babylon there would be more stars with a circumpolar path and a more marked southward tilt to the paths.

doubt that we could complete the paths of the northern circumpolar stars in the sky and the paths of the remaining stars below the horizon by drawing complete circles. This is an extrapolation from our observations and constitutes about the simplest and most plausible kind of theorizing there is.

Repeated observations over a fairly long period would reveal that apart from the Sun, the Moon, and five other bodies—Mercury, Venus, Mars, Jupiter and Saturn—each of the stars with a path cutting the horizon rises and sets at the same point on the eastern and western horizons but at a slightly earlier time night by night (or by extrapolation, day by day). The Ancients regarded all seven exceptions to the general rule just stated as planets (Greek *planetes*, a wanderer). In midsummer on the longest day of the year in Babylon, the Sun rises about 28 degrees north of east and sets about 28 degrees north of west. On succeeding days it rises and sets at a less northly point until at the autumn equinox (when the day and the night are equal in duration) it rises due east and sets due west. As autumn passes into winter, its position moves further south until at midwinter on the shortest day of the year it rises and sets about 28 degrees south of east and of west respectively. The southern limit of the Sun's movement is termed the Winter Solstice, the winter day when the Sun stands instead of moving further south. Thereafter the Sun begins its journey northward back to the Summer Solstice, where once again it pauses and then begins its southward journey. The daily path of the Sun across the sky can be seen from Babylon always to be tilted back, like the paths of the fixed stars, 32.5 degrees towards the south (the latitude of Babylon is 32.5 degrees north). At midday in midsummer the Sun can be seen to be only 8 degrees from the zenith, whereas at midday in midwinter it can be seen to be 56 degrees from the zenith, that is, over half way towards the southern horizon.

The full cycle of the Sun's movement towards the south and back to the north may be seen to occupy, in round Babylonian numbers, some 360 days. The precise day on which the Solstice occurs is not easy to determine as that day differs so little in length from its immediate neighbours and the Sun's deviation north or south is little different from what it is on the day or two on either side. Babylonian devices for the measurement of time and direction were not equal to a precise determination. However, after many centuries of trying to keep the annual calendar and the seasons in step, it became clear, just as it had to the Egyptians as early as 4000 BC, that the year of four seasons was about 365 days or more precisely $365\frac{1}{4}$ days. This period is called the tropical year from the Greek word *trope*, a turning point, as it is marked out by the turning south (or north) of the Sun.

The Babylonians chose a more easily determined period as the main unit of their calendar. It is that defined by the change of shape of the Moon rounded into days. The Sun and the Moon are sometimes in conjunction, that is, abreast with one another in their journey across the sky, sometimes in opposition, that is, half the circle of the heavens apart. When they are in conjunction, the Moon is invisible but within a day or so it is seen as a thin crescent. When they are in opposition, the Moon is a full disc. When they are at intermediate degrees of separation, the Moon passes through quarter, half, and gibbous phases. This variation in angular separation involves the Moon falling behind the Sun about 12 degrees day by day in their apparent westward journey. In $29\frac{1}{2}$ days the Sun catches up with the Moon, or laps it as runners say. This period is called the synodic month, the period between successive 'meetings' in the sky of the two bodies (*sunodos* is the Greek word for a meeting). Twelve such moon-periods equal 354 days, only 11¼ days short of the tropical year. After some centuries the Babylonians adopted a calendar with most years having twelve months interspersed with occasional years having thirteen months.

Very careful observation and record keeping reveal that the fixed stars are moving a little more swiftly westward than the Sun which we have seen is swifter in its westward journey than the Moon. One way in which a Babylonian observer could observe this involved records of the heliacal rising of some prominent fixed star. An heliacal rising is one occurring just before sunrise (*helios* is the Greek for the Sun). Ten days after its heliacal rising a star may be seen to rise some 40 minutes before sunrise and to be almost 10 degrees above the horizon at sunrise; twenty days later it would rise some 80 minutes before sunrise; after about 365 days, or 365¼ days on the average over a few years, the star has its next heliacal rising. That is, in 365¼ days the star has lapped the Sun in its westward journey around the sky over and over again.

The Sun is thus not fixed relative to the stars but wanders eastward in a year through them. In the previous paragraph it was said that a fixed star lapped the Sun in its westward journey; it is now being said that the Sun wanders eastward through the stars. These are alternative ways of saying the same thing. Imagine a tide running strongly from A to B, with a free floating plank and a boat being carried along by it. Should the boat be driving into the tide, one may think of the closing distance between the plank and the boat in either of two ways, namely the plank overtaking the boat or the boat approaching the plank. If we think of the stars as floating in a westward running tide then the Sun drives eastward against the tide that carries it westward too.

The Sun's eastward drive through the westward moving stars is along a path which over a year is to-and-fro or zig-zag in pattern. This path cuts the celestial equator, an imaginary circle cutting the horizon due east and due west (and above the horizon at Babylon tilted 32.5 degrees towards the south), twice at an angle of 23.5 degrees as shown in Fig. 3.2. This path is called the Ecliptic for a reason we shall later discover. The path of the Moon (also eastward through the stars) and the paths of the other five wanderers (predominantly eastward with an occasional westward retrogression) weave to-and-fro

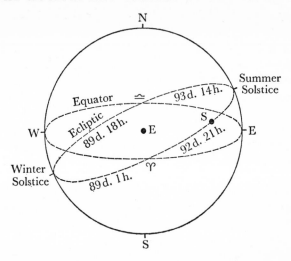

Fig. 3.2 The circle represents the apparent sphere of the sky. N and S are the north and south celestial poles respectively. The Equator is shown as a plane perpendicular to the axis of the poles. The Ecliptic is shown as a plane cutting the equatorial plane at an angle of 23.5 degrees. The former cuts the latter at the points marked with the symbol of the ram (the Spring Equinox) and of the balance (the Autumnal Equinox). The time taken by the Sun to traverse each quarter of the Ecliptic is shown.

across the Ecliptic but within a band 18 degrees wide. This band, which was called by the Greeks the Zodiac or the Circle of the Beasts, was divided by the Babylonians into twelve equal segments, each spanning 30 degrees. Each segment or 'sign' was occupied by a recognized constellation or pattern of stars and was given a name, usually that of an animal, for example, the bull, the ram, the fishes, the goat, and the lion. Six signs divide the Sun's path north of the celestial equator and six its path to the south.

Careful observation over a long period reveals that the Sun moves along the Ecliptic with a variable pace. The journey north of the equator takes more time than the southern movement. It takes 92 days 21 hours from the northward crossing of the Equator to the top of its northern swing and 93 days 14 hours back to the Equator, whereas it takes 89 days 18 hours from the southward crossing to the bottom of the southern swing and 89 days 1 hour back to the Equator again. That is, the period spent by the Sun north of the celestial Equator is 7 days 16 hours longer than that spent south of it. This change of pace came to be spoken of as an anomaly, that is, not according to the law (from the Greek *nomos*, a law). The Greek theoreticians assumed that celestial bodies moved, or ought to move, with uniform velocity.

The Moon, too, has an anomalistic change of pace, the period between successive occasions of maximum angular velocity being $27\frac{1}{2}$ days. This period is the anomalistic month. There are two other readily established moon periods. While the Moon takes $29\frac{1}{2}$ days to come into conjunction with the Sun again, it takes only $27\frac{1}{3}$ days to come again into conjunction with any fixed star. The shorter period is the sidereal month. The Moon zig-zags across the Ecliptic (the Sun's path) approximately twice each month. These crossings are called the nodes of the Moon, the northward crossing being the ascending node and the southward crossing being the descending node. The period between corresponding nodes, e.g. successive ascending nodes, is $27\frac{1}{5}$ days. This is called the draconic month (from the Greek *dracon*, a dragon or serpent). Eclipses occur only when the Moon is at or near a node and the Sun is at that node or the opposite node. Early superstition attributed the darkening of the Moon or Sun, as the case might be, to its being swallowed by a celestial dragon.

It may be instructive to deviate here from a further description of the apparent motions of the celestial bodies in order to comment on the Babylonian number system. Our present system is decimal (Latin *decimus*, tenth) as we count in units from 1 to 10 and then begin another cycle of ten units to arrive at 20 (i.e. 2 × 10) and so on. The Babylonian system was sexigesimal (Latin *sexigesimus*, sixtieth), as they counted in units from 1 to 60 and then began another cycle of sixty units to arrive at 120 (i.e. 2 × 60) and so on. Thus the numerals 13 mean for us one ten plus three units. The Babylonian analogues meant one sixty plus three units, i.e. what 63 means to us. It is from the Babylonians that we inherit our division of the hour of time and the degree of angle into 60 minutes, and the minute of time and of angle into 60 seconds. The number 360, or 6 times 60, was for the Babylonians a major round number. It was near enough to the number of days in the year of four seasons for the year to be taken to be 360 days. It was natural enough for the Babylonians to say that the Sun moved one degree of angle each day and so divide the circle into 360 degrees as we still do. It is said that some Chinese astronomers, possibly under Babylonian influence but knowing the year to be 365¼ days, divided the circles on their instruments into 365¼ degrees. With a synodic month, that is 29½ days, it was convenient for the Babylonians to round the month into 30 days (i.e. a half of 60), and in turn to divide the year of 360 days into twelve months of 30 days each. Twelve, like 30, is a simple fraction of 60. It is not surprising that the Babylonians divided the average day and the average night each into twelve hours and divided the Zodiac into twelve signs each spanning 30 degrees.

Just as there are several months of slightly different period so are there several years though the Babylonians seem not to have distinguished them. Hipparchos, a Greek astronomer of the second century BC, showed that the sidereal year is slightly longer (a little over twenty minutes is the modern reckoning) than the tropical year. The latter can be determined by taking an average over many years of the period between Summer Solstices taken as the day on which a gnomon casts the shortest shadow. The former can be determined by taking over many years the period between successive heliacal risings of a given fixed star. The sidereal year is analogous to the sidereal month, and the tropical year to the draconic month. There is also an anomalistic year but it is less than five minutes longer than the sidereal year. Hipparchos recognized that the difference in period between the sidereal and the tropical year implied that the equinoxes, the points where the Equator cuts the Ecliptic, were slowly moving westward (precessing) along the Ecliptic. He gave the value as 40 seconds of angle per year, a fair approximation to the currently accepted value of 50.3 seconds. Thus in his time the Spring Equinox occurred at the first point of the ram (we still use the symbol ♈, presumably the ram's horns, for the Spring Equinox), whereas today it occurs in the next sign, the fishes. About 2400 BC it would have occurred at the first point of the bull. The Babylonians often called this sign the 'Bull in Front', that is, the animal which led in the year—until modern times the year was deemed to begin at the Spring Equinox. This suggests that the signs of the Zodiac go back in Babylonian history to the third millennium BC. In Hipparchos' day the bull was second in the year's procession of the beasts. Today he is third. Thus, though slowly lagging, he is also slowly making his way again to the front of the Grand Parade. The precession of the equinoxes not only shifts the

Sun at the time of the Spring Equinox back one sign after about two millennia but also shifts the position of the north celestial pole in the field of the fixed stars.

There remains for us the plotting of the courses of the five other wanderers. For convenience we will take Mercury and Venus first. The latter is the major evening and morning star to be seen shining brightly, sometimes in the western sky after sunset and sometimes in the eastern sky before sunrise. It is never far away from the Sun, sometimes disappearing in the dazzling light near to the Sun's disc. Mercury too always stays close to the Sun, indeed so much closer than Venus that it is not easily seen. The two are like dogs held by the Sun on leashes—Venus on a longer leash than Mercury. Sometimes they run ahead of the Sun on its eastward journey through the fixed stars, sometimes they fall behind, indeed they run back westward amongst the stars. Venus' leash allows her to lag or to run ahead of the Sun by as much as 48 degrees. Mercury's leash sometimes stretches out to 28 degrees but sometimes it pulls back the scampering pup from a point separated by only 18 degrees from the Sun. When these planets come level with the Sun as they overtake it on its eastward journey, they are, in modern terms, in superior conjunction. This happens at intervals of 116 days for Mercury and 584 days for Venus. These are their synodic periods, the intervals between their successive overtakings of the Sun. At the middle of their retrograde phase they come into inferior conjunction. As they are both 'tied' to the Sun, they both take $365\frac{1}{4}$ days to make a complete circuit of the Zodiac. That is, the two have the same zodiacal or sidereal period.

The other three wander far away from the Sun, coming into opposition with it from time to time. They too have a retrograde phase (their opposition with the Sun occurs midway through it) though their usual motion through the stars is direct (eastward). Their synodic periods, that is, the intervals between successive conjunctions with the Sun, are as follows: Mars, 2 years 49.5 days; Jupiter, 1 year 34 days; Saturn, 1 year 13 days. Overall they steadily fall behind the Sun, the conjunctions occurring as the Sun laps them. The intervals of their return to a given point in the Zodiac are quite different from their synodic periods. Because it will be more relevant to later parts of the story, the periods of return to a given point in the Zodiac as seen by an imaginary observer on the Sun, rather than by an actual observer on the Earth, will be given. They are for Mars, 1 year $321\frac{3}{4}$ days, for

Jupiter 11 years 315 days and for Saturn, 29 years 167 days. These are their zodiacal or sidereal periods.

All five follow paths which cross back and forth over the Ecliptic. Venus strays furthest, marking the edge of the Zodiac in sometimes deviating from the Ecliptic by 9 degrees. Mars is next with deviations of 7 degrees, followed by Mercury with deviations of 5 degrees and Jupiter and Saturn, each with maximum deviations of less than 3 degrees. The combination of alternating direct and retrograde motion with these deviations from the Ecliptic produces looped paths like that illustrated in Fig. 3.3.

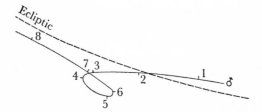

Fig. 3.3 The path of Mars over a seven-month period in 1939. The numbers indicate its position on the first day of the successive months. Its motion is retrograde from position 4 to position 6, and direct from positions 1 to 4 and from positions 6 to 8.

The Babylonians believed that the seven wanderers took it in turn to dominate successive days in the following order: Sun, Moon, Mars, Mercury, Jupiter, Venus, and Saturn. Each day bore the title of its guardian, a usage preserved in Latin and showing obvious remnants in many contemporary Western languages. Thus if we mix a little French and English, we have Sunday, Monday, *mardi*, *mercredi*, *jeudi*, *vendredi*, and Saturday. The seven day week of the Babylonians provided a smaller unit of time than the year and the month. It was, unfortunately, not a simple fraction of either, just as the month was not strictly a simple fraction of the year. The sidereal year and the synodic month are 'natural' units of time, whereas the week is arbitrary. In order to have a year and a month with a round number of days, we adopt an arbitrary year of 365 days, or in leap years 366 days, and arbitrary months of 28 or 29 or 30 or 31 days.

Apart from these arbitrary periods, practically all the information given above consists of facts of observation. Few of these facts are of the sort one can glean in a single act of observation. We can see that this evening the Moon is crescent and we can see less than fourteen days later that it is full. Only by comparing these and a number of intervening observations can we say that the Moon changes progressively from crescent to full. We have to watch a number of cycles of lunar changes and strike an average in order to say that the synodic period and the length of the cycle (a lunation) are the same and are $29\frac{1}{2}$ days. We cannot watch the Moon through every minute step in its changes as it is sometimes out of sight below the horizon. But we are disinclined to believe that the change in shape between moonset and moonrise has been achieved in one sudden jump. We believe the kind of change that occurs as we watch is continued in the change that occurs when we cannot see it. This is, of course, an assumption. It is the very simple sort of theorizing which we have already seen in the suppositions that the circumpolar stars complete in daylight the circles begun at night and that the celestial bodies, which set and rise, continue to move below the Earth on the curved paths we see them following above the Earth.

Though some of our facts are 'brute' facts in so far as they lack an apparent reason and are not dependent upon anything else about which we have claimed knowledge, others are clearly dependent upon other facts which we know. Why the Sun takes $365\frac{1}{4}$ days and the Moon $27\frac{1}{3}$ days to circle the sky is so far a mystery to us. They are without any obvious reason. However, given them, it may be seen that the synodic month must be $29\frac{1}{2}$ days. Starting from a given meeting, the Moon makes one circuit of 360 degrees in $27\frac{1}{3}$ days. In this time the Sun travels $27\frac{1}{3} \div 365\frac{1}{4} \times 360°$, or almost 27 degrees. In another $2\frac{1}{8}$ days, that is, $29\frac{1}{2}$ days from the start, both the Moon and the Sun move on to a point almost 29 degrees from the starting point. Thus the synodic month of $29\frac{1}{2}$ days is dependent upon the sidereal month of $27\frac{1}{3}$ days and the sidereal year of $365\frac{1}{4}$ days. Were we really Babylonians, a period of 360 days for the Sun and $27\frac{2}{3}$ days for the Moon would seem less peculiar for they would yield a synodic month of 30 days, just twelve of which would fit in a year. Such a neatness could leave us with an impression of rightness which is often taken as the mark of the guiding hand of some superior Reason at work.

The year and the sidereal period of any planet similarly determine the synodic period of that planet. The relatedness of these facts enable predictions if one makes the assumption that they will continue in the future as we have found them to be in the past. A more elaborate piece of prediction was made by the Babylonian astronomers in connection with eclipses. For an eclipse to occur the Moon must be at or near a node and the Sun must be either there too (in which case there is an eclipse of the Sun) or at the opposite point on the Ecliptic (in which case there is an eclipse of the Moon), granted that the conjunction occurs in the daytime and that the opposition occurs at night. Given these facts and the facts already noted about the periods of the Sun and of the Moon, and about the regression of the Moon's nodes around the Ecliptic, it can be worked out that eclipses occur in a pattern which repeats itself every 18 years 11 days. This period is the *Saros*. It was discovered apparently as an independent fact by the Babylonians, though it can be deduced from the related facts. Clearly it may be used to predict eclipses even though one does not understand that an eclipse of the Sun is caused by the Moon obscuring it and an eclipse of the Moon by the Earth's shadow. One does not necessarily have to understand in order to predict.

In order to make some sense of the diurnal motion of the fixed stars in solid formation one could be tempted to say that they looked as though they were stuck on to (or were perhaps holes in) a dark sphere which rotated east to west in 24 hours on an axis through the North Pole and an analogous (but imagined) South Pole 32.5 degrees below the southern Babylonian horizon. In a way, it does look as though that is the case, but nevertheless any contention that it is the case is a supposition and not an observation.

Eudoxos' theory of concentric spheres

Our supposition about the celestial sphere bearing the stars was made by Eudoxos, a Greek living in the first half of the fourth century BC. He spoke of the sphere itself as invisible, as a crystalline sphere. It is hard to tell whether he meant this to be taken literally or whether he was merely speaking metaphorically as a geometrician. We do not have his own account of the matter; there are only the accounts given by later writers some of whom clearly got some details of his whole theory wrong. It is possible, of course, that he was not sophisticated enough in the philosophy of science to make the distinction between a literal and a metaphorical statement about a scientific matter. The following account will be written as though he was speaking literally,

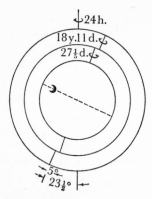

Fig. 3.4 Eudoxos' model for the Moon's motion.

although it has been customary to treat his theory as mathematical rather than mechanical.

While one sphere serves him to account for the apparent motions of the fixed stars, Eudoxos calls upon three concentric spheres shown schematically in Fig. 3.4, to account for the apparent motions of the Moon. The outer sphere, not only invisible but also starless, is assumed to rotate east to west in 24 hours on a polar axis. Indeed the sphere of the fixed stars could be substituted for it. The middle sphere rotates east to west in 18 years 11 days (223 lunations, or the *Saros*) on an axis fixed to the outer sphere at right angles to the Ecliptic (or 23½ degrees to the polar axis); it too is invisible and starless. The inner sphere rotates west to east in 27⅕ days (the draconic month) on an axis fixed to the middle sphere and at some 5 degrees to the axis of that sphere. The Moon is borne on the equator of the inner sphere. The rotation of the outer sphere accounts for the diurnal motion of the Moon. The rotation of the middle sphere accounts for the regression of the Moon's nodes and for the Moon's fairly close adherence to the Ecliptic. The rotation of the inner sphere accounts for the draconic month and the Moon's deviation of up to 5 degrees from the Ecliptic. The rotations of the middle and the inner spheres taken together account for the period of the sidereal month, 27⅓ days, the extra fraction of a day being added to the 27⅕ days of the inner sphere by the very slow rotation of the middle sphere in the opposite direction.

We may say that the mechanism accounts for the apparent motions because it implies or requires them. Were there indeed a mechanism as described then the several effects mentioned in the last paragraph could only be as they are; they could not be otherwise and to that

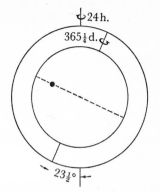

Fig. 3.5 A simplified version of Eudoxos' model for the Sun's motion.

extent they are explained. Eudoxos' mechanism for the Moon, however, does not explain the Moon's anomalistic changes of pace. Instead it implies that the Moon's angular velocity is uniform. To this extent the theory is inadequate.

Eudoxos' mechanism for the Sun also uses three spheres, the middle one being introduced to account for an effect that is not observed. Many of the Ancients assumed that because the other six wanderers strayed from the Ecliptic, the Sun must do so as well. In the following account we shall confine ourselves to the two needed spheres (*see* Fig. 3.5). The outer one, once again, is a starless sphere rotating in 24 hours east to west on a polar axis. The inner one rotates west to east in 365¼ days on an axis set at 23.5 degrees to the axis of the outer sphere. The Sun is borne on the equator of the inner sphere. The mechanism accounts for the north-south cycle of the Sun in a year as well as accounting for the length of the year. It does not, of course, account for the anomalistic change of angular velocity (a fact that Eudoxos may have known) nor for the difference in period between the tropical and the sidereal year (a fact not known by Eudoxos). The change of pace could be taken care of by assuming that the inner sphere rotates not with uniform but with varying angular velocity. Eudoxos may have been reluctant to make this assumption probably because he accepted a tradition, first broken by Kepler in the seventeenth century AD, that the celestial motions must be at basis circular and of uniform velocity.

The full ingenuity of Eudoxos' analyses of apparent celestial motion can be seen in his mechanisms for the five planets. The outer sphere, once again, is starless and rotates in 24 hours east to west on the polar

29

axis. The next sphere is also starless and rotates in the zodiacal period of the planet west to east on the polar axis. The first sphere accounts for the diurnal motion of the planet and the second for its general direct motion around the Zodiac. Once again he makes no attempt to account for the anomaly of this overall motion. The third sphere, also starless, rotates, on an axis in the plane of the Ecliptic, in the synodic period of the planet. The inner sphere, bearing the planet on its equator, rotates in the synodic period also but in a direction opposite to that of the third sphere; its axis being inclined to that of the third by an amount that varies from one planet to another. The joint action of the last two spheres produces an alternating looped backwards-and-forwards motion or figure of eight. When this is combined with the motion of the second sphere one loop is drawn out to give the long phase of direct motion and the other compressed to give the short phase of retrograde motion. The angle between the axes of the third and fourth spheres is set to give the maximum deviation of the planet from the Ecliptic. The axes of the third spheres for Mercury and Venus pass through the same points of the Ecliptic, whereas those for the other three have other locations; this provides the basic distinction between the motions of these two sets of planets.

That Eudoxos started each mechanism afresh with an outer sphere suggests that he recognized that it was not possible to put the several mechanisms together—the spheres of one would obstruct the eccentrically placed axes of another. Nevertheless to account for some facts such as eclipses and the length of the synodic month and of the synodic periods of the planets, he would have to think of them together. Still if a sphere can be regarded as invisible there is probably no serious objection to regarding it as offering no resistance to an axis ploughing through it.

Despite its great ingenuity the theory did not account for all the facts. Eudoxos may be forgiven for not accounting for phenomena about which he did not know. On the other hand, it would have been a great triumph for the theory were it able to predict what later proved to be facts though not known to be such by its author. We have already noticed that at one point the theory went wrong in accounting for something that was thought to occur but which does not occur. Because the other wanderers stray from the Ecliptic it was thought proper that the Sun should too, even if the amount were imperceptible. Science is often hindered by what some scientist thinks ought to be or must be. From Eudoxos in the fourth century BC to Copernicus in the sixteenth century AD nobody seriously questioned the assumption that the real celestial motions are circular and of uniform velocity. The problem was taken to be that of accounting for the apparent nonconformity of the wanderers in one or both of these respects. Nobody was ready to argue that the planets were in fact non-conformists. Broad principles like these are all too likely to be used as Procrustes used his bed. Facts that are too long are likely to be lopped and those too short to be stretched. The contention that nothing can have a velocity in excess of that of light is sometimes enunciated in elementary textbooks as though it were some isolated necessity standing on its own base or a completely established fact of observation. What we should recognize is that modern physical theory supported by a great deal of observation does not provide a bed long enough to accommodate any 'fact' about entities with velocities greater than the velocity of light. Were such a 'fact' to turn up we would probably want to chop its legs off. However were we convinced of its truth we would set about building a new bed.

Part of the ingenuity of Eudoxos' theory consists of the way in which axes are so offset or periods of rotation are so fixed that they give the desired results. In this respect they are *ad hoc*, that is, they are chosen in order to give the desired results. But they do not, unfortunately, lead to other independent results. The theory implies nothing that is not already explicit in the facts it is meant to account for. For this reason it is a sterile theory. As we shall see later it is a theory of a sort which, if revision by extension or by change of the *ad hoc* values is allowed, is untestable. The full meaning of this will be left, however, to later discussion.

There is another aspect of the criticism of *ad hoc* theorizing. To say that the stars are fixed to a sphere rotating east to west in 24 hours on a polar axis and that the Sun is fixed to the equator of an invisible sphere rotating west to east in 365¼ days on an axis carried by the outer sphere but placed at an angle of 23.5 degrees to its own axis gives some kind of account of the observed facts but it is a limited account. In a way it only passes the problem back. It is said that a Hindu philosopher suggested that the world was supported by an elephant. But in order to support the elephant he placed it on the back of a giant tortoise which was supported by the sea in which it swam. To bring an end to this successive shifting of the problem, he asserted that the sea was bottomless. He might as well have admitted at the beginning that he did not know how the world was supported.

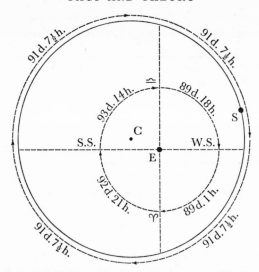

Fig. 3.6 Hipparchos' explanation of the anomaly (change of pace) of the Sun in terms of an eccentric. The Sun follows a circular path centred on C with uniform angular velocity. As seen from E, the Earth, the Sun takes different times to traverse the four quarters of the Ecliptic. The symbols of the ram and of the balance mark the spring and autumn equinoxes respectively, and SS and WS the summer and winter solstices respectively.

Finally, it must be pointed out that even if Eudoxos' mechanisms are taken literally his theory says nothing about the engines driving the nests of spheres; it says nothing about the dynamics of the celestial motions.

Ptolemaic theory

The bases for an alternative to the theory of Eudoxos were adopted in the second century BC by Hipparchos, though their full exploitation awaited Ptolemy in the second century AD. These two astronomers were the pioneers of star cataloguing; Ptolemy's catalogue, which was based upon Hipparchos', has been preserved for us in his book, named by the Arabs, *Almagest*.

Hipparchos made use of two geometrical devices, the eccentric and the deferent-epicycle system. They can account in general terms for the anomalous changes in apparent angular velocity, with reference to an observer on the Earth, of the seven wanderers. The account nevertheless preserves the belief in real circular motions of uniform velocity.

The principle of the eccentric may be illustrated by means of one of Hipparchos' alternative theories of the apparent motion of the Sun. Suppose the Sun moves with uniform velocity around the circumference of a circle whose centre is offset from the position of the Earth. Then, when the Sun is moving on that part of its orbit lying about the point nearest the Earth (*perigee*) it will have, to an observer on Earth, a greater apparent velocity than when it is moving on that part of its orbit lying about the point furthest from the Earth (*apogee*). This principle is represented in Fig. 3.6. It will be noted that the amount by which the Earth is offset from the centre of the Sun's orbit is *ad hoc* that is, it is fixed in order to give the required anomaly.

The principle involving a deferent and an epicycle may best be expounded in purely geometrical terms first and subsequently given an astronomical application. One circle, the epicycle, is deemed to have its centre moving around the circumference of a second circle, the deferent. A point on the circumference of the epicycle may rotate in the same direction as that in which the centre of the epicycle moves around the deferent, or it may rotate in the opposite direction. The point on the epicycle may make one rotation while the centre of the epicycle makes one, or it may make some fraction or some multiple of one rotation while the centre of the epicycle makes one rotation. Case A in Fig. 3.7 is one where the motions are opposite and where, according to the method of reckoning adopted by the Ancients, the periods of rotation are equal. It will be noted that the figure traced by the moving point on the epicycle is an eccentric in respect of a body at the centre of the deferent. Thus, as Hipparchos recognized, this provides an alternative mathematical expression of his eccentric theory of the Sun's motion. Case B is one where a point on the epicycle traces an ellipse, the motions being opposite and the period of the epicycle being half that of the deferent.

The Ancients reckoned a rotation of the epicycle relative to a moving radius of the deferent and not, as we now regard as more natural, relative to a fixed external framework. Thus were a rigid bar bolted in a fixed position at the edge of a disc, the Ancients would have said that it made no rotations as the disc rotated, whereas we would say it made one direct rotation for each direct rotation made by the disc.

33

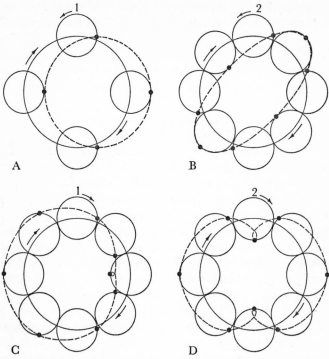

Fig. 3.7 The generation of various curved figures by means of epicycle and deferent. In each case the rotation of the deferent is clockwise. In A and B the rotation of the epicycle is anti-clockwise and in C and D it is clockwise. In A and C, the epicycle and the deferent have equal periods as the Ancients reckoned them and in B and D, the epicycle has a period half that of the deferent.

Thus, in the modern reckoning, case A shows no rotation of the epicycle while the deferent rotates once, and case B shows one retrograde rotation of the epicycle during one direct rotation of the deferent.

Case C shows one (Ancient) direct rotation of the epicycle during one direct rotation of the deferent. The figure traced by the point on the epicycle is a closed curve. When as in the example depicted the ratio of the radius of the epicycle to the radius of the deferent exceeds a certain value, the curve is looped. The maximum angular velocity of the point for an observer at the centre of the deferent occurs when the point is furthest from the centre, that is when the direct motions of both circles are combined. When the point on the epicycle is nearest

34

the observer the motion of the point on the epicycle due to the rotation of the epicycle is opposite to that due to the motion of the deferent; there may be, as in the present case, a resultant brief phase of retrograde motion of the point on the epicycle. Case D shows the way in which two loops are produced in the curved figure traced by the point on the epicycle. It can be readily seen that the number of loops, if they occur, equals the ratio of the period of the deferent to the period of the epicycle (in the Ancient reckoning). Further as the ratio of the radius of the epicycle to the radius of the deferent increases there is an increase in the variation of angular velocity as seen from the centre of the deferent and an increase in the extent of the phase of retrograde motion.

Hipparchos used the epicycle-deferent system in his theory of the Moon. He set the deferent, centred on the Earth, with an inclination of 5 degrees to the plane of the eccentric bearing the Sun. The nodes, the points where the Moon's deferent cut the Ecliptic, were assumed to move slowly backwards (i.e. East to West) around the Ecliptic in a period of $18\frac{2}{3}$ years. The epicycle, on whose circumference the Moon is borne, was assumed to rotate in the retrograde direction while the deferent rotated directly (i.e. West to East). The period of the epicycle was taken to be $27\frac{1}{2}$ days (the anomalistic month) and that of the deferent $27\frac{1}{3}$ days (the sidereal month); the radii of the epicycle and of the deferent were taken to be in the proportion $21:240$, insufficient to yield a retrograde phase.

The several features of the theory enable a reasonable explanation of the facts as reported above. Thus the deferent being tilted 5 degrees to the Ecliptic, the path of the Moon will deviate above and below that of the Sun by up to 5 degrees. The cycle of changing angular velocities yielding the anomalistic month is accounted for by the retrograde rotation of the epicycle in $27\frac{1}{2}$ days. When the Moon is on the 'inside' segment of the epicycle (and thus moving in the same direction as the deferent), it will have maximum velocity. When it is on the 'outside' of the epicycle (and thus moving in the opposite direction to the deferent), it will have minimum velocity. The retrogression of the nodes is attributed to a slow 'wobble' in the plane of the deferent relative to the Ecliptic. And so on. Hipparchos himself recognized that though the theory gave a good account of the position in the sky of the Moon at conjunction with and opposition to the Sun, it misplaced the Moon a little on some occasions of quadrature (i.e. when the Moon was in its first or its third quarter). Ptolemy later

showed how this defect could be remedied by assuming the deferent
to be centred not on the Earth but on a point displaced from it.
However, it was necessary for him to assume that the deferent moved
with uniform angular velocity at the Earth and not at its own centre.
In making this last assumption Ptolemy was in reality abandoning
the principle of uniform velocity, though seeming to preserve it. He
made another departure when he used the notion of the equant
(*punctum equans*) in his planetary theories. This, as Fig. 3.8 shows, may
be used to account for anomalous variations in apparent velocity.

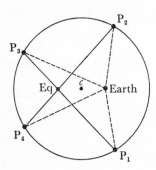

Fig. 3.8 The planet P follows a
circular orbit centred on *c*. Its angular
velocity relative to Eq, the equant, is
uniform. As seen from Earth, its
angular velocity undergoes marked
changes, so that the arc $P_1 P_2$ of 160
degrees is traversed in a quarter of
the planet's sidereal period.

Hipparchos considered his data on the other five 'wanderers'
inadequate for the formulation of any precise theories for them. With
better data, gathered in the intervening period at his disposal, Ptolemy
provided these theories. Their broad features are as follows:

(i) The deferent of each planet is eccentric to the Earth to an
extent that is sufficient to account in part for the overall inequality of
the planet's motion around the Zodiac in its zodiacal period.

(ii) The deferent of each planet is inclined to the plane of the
Ecliptic to a degree that is sufficient to account for the maximum
deviation of the planet's path from the Ecliptic.

(iii) The rotation of the deferent of each planet is direct, i.e. West
to East.

(iv) The angular velocity of a point on the circumference of the
deferent is uniform neither at the centre of the deferent nor at the
Earth (eccentrically placed) but at the equant, a point equidistant
with the Earth from the centre but on the side of the centre directly
opposite the Earth.

(v) The periods of the deferents of Mercury and Venus are equal,

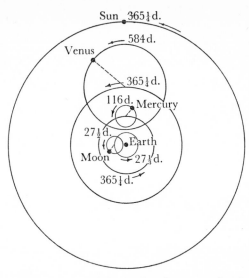

Fig. 3.9   Basic features of the Ptolemaic system for the Moon, Mercury, Venus, and the Sun. NOTE that the centres of the epicycles of Mercury and Venus lie on the radius vector from the Earth to the Sun. No significance should be attached to the overlap of the epicycles of the Moon, Mercury, and Venus; it has been introduced in the diagram for convenience of drawing.

being a sidereal year, i.e. 365¼ days. The periods of the deferents of Mars, Jupiter and Saturn are the zodiacal periods of those planets, namely 687 days for Mars, 11.86 years for Jupiter and 29.46 years for Saturn.

(vi) Not only do the deferents of Mercury and Venus have the same period of rotation as the Sun, but also the point on the circumference where the centre of the epicycle lies is on the radius vector from the Earth to the Sun.

(vii) The epicycle of each planet also rotates directly, the angular velocity being uniform at its centre.

(viii) The period of each epicycle, reckoned in the Ancient manner, is the synodic period of the planet, that is, 116 days for Mercury, 584 days for Venus, 780 days for Mars, 399 days for Jupiter and 378 days for Saturn. Had these periods been reckoned in the modern manner they would have been 88 days for Mercury, 225 days for Venus and 365¼ days each for the other three.

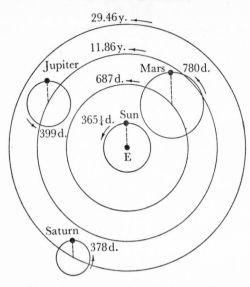

29.46 y.

11.86 y.

Jupiter

Mars   780 d.

687 d.

399 d.   365¼ d.   Sun

E

Saturn   378 d.

Fig. 3.10 Basic features of the Ptolemaic system for the Sun, Mars, Jupiter, and Saturn. NOTE that the radius vector from the Earth to the Sun and the radius vectors from the centre of each epicycle to the planet are parallel. No significance should be attached to the overlap of the epicycles.

(ix) The radius vector from the centre of the epicycle to the planet in the cases of Mars, Jupiter, and Saturn remains parallel to the radius vector from the Earth to the Sun.

(x) The ratio of the radii of epicycle and deferent is 0.3708 for Mercury, 0.7194 for Venus, 0.6583 for Mars, 0.1917 for Jupiter and 0.1083 for Saturn.

This is a quite complex geometrical theory. Ignoring the eccentricity of the deferents with respect to the Earth and the inclination of their planes to the plane of the Ecliptic, the main features of the set of theories may be represented as in Figs 3.9 and 3.10. Ptolemy put the seven wandering bodies in the order Moon, Mercury, Venus, Sun, Mars, Jupiter and Saturn from the Earth as shown. However, he did not require their relative distances to take any determinate values, hence no heed should be given to the spacing of the seven systems in these diagrams.

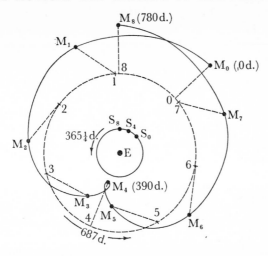

Fig. 3.11   The basic features of the Ptolemaic theory of Mars. Beginning at time 0, the Sun and Mars are in conjunction. 390 days later at time 4, they are in opposition, the Sun having made about one and a fifteenth circuits and Mars a little over half a circuit of the Zodiac. 780 days later at time 8, they are in conjunction again, the Sun having made about two and a seventh circuits and Mars about one and a seventh.

It can readily be seen why, in terms of this theory, Mercury and Venus never stray far from the Sun and why the other three planets as well as the Moon can move into opposition with the Sun. The synodic and zodiacal periods of the five planets can be readily deduced from the theory. The reasons for the occurrence of maximum forward motion when the planet is in conjunction and of maximum retrograde motion when it is in opposition, in the cases of Mars, Jupiter and Saturn are also provided. And so on for most, if not all, of the other observed facts of planetary motions as reported above.

In order to see the theory of one planet in some detail, let us consider the case of Mars. In Fig. 3.11 the deferent and the epicycle are drawn with radii in the proportion 1 : 0.66. Beginning with the centre of the epicycle at 0 and Mars on its circumference at M_0 the Sun being, for convenience in this diagram, at S_0. After 195 days, the centre of the epicycle will have moved approximately 0.29 (195/687)

39

anti-clockwise around the deferent from 0 to 2 and Mars will have moved 0.25 (195/780) anti-clockwise around the epicycle to M_2. After 390 days, the centre of the epicycle will have moved to 4 and Mars to M_4. At this point Mars will be midway through its short period of retrograde motion. It will be seen that the Sun beginning at S_0 when Mars was at M_0, i.e. Mars was in conjunction with the Sun, will in $365\frac{1}{4}$ days have moved back to that point. In another $24\frac{3}{4}$ days (i.e. in a total of 390 days), it will have completed almost 0.07 of its second time around and will be at S_4. That is, Mars while in the middle of its retrograde phase will be in opposition to the Sun. When Mars has completed a full turn around the epicycle (in 780 days), the centre of the epicycle will be almost back to position 1 and Mars will be at M_8, and in conjunction with the Sun at S_8 (the Sun having made about 2.13 turns around its orbit). Thus one synodic period has been completed.

The theory is one of very great geometrical ingenuity. Some of its defects can not be brought out effectively until we have considered Copernicus' heliocentric theory, especially as refined by Kepler. However, one point may be made now. Like Eudoxos' theory, all the values used—inclinations of deferents, periods of deferents and of epicycles, degrees of eccentricity, ratios of the radii of epicycle and deferent—are *ad hoc*. They are chosen so that they will give the desired result in some particular case and they lead to no conclusions, or at any rate no precise conclusions, other than those concerned with just this case. If an insufficient set of these facts were considered in the determination of these *ad hoc* assumptions, as happened in Hipparchos' theory of the Moon, then some other *ad hoc* assumption can be made as Ptolemy did when he made the deferent eccentric. Had this not sufficed then a second epicycle could have been superimposed on the primary epicycle, as was done by many of Ptolemy's followers in other cases. Or an equant could have been introduced or the radii of the deferent and of the epicycle changed or some other variation not beyond human devising made. The theory is so protean that it is almost beyond attack—attack it while it is one of its shapes and it will elude defeat by changing into another. Granted that apparent size or apparent brightness is a function of distance, the theory had some implications for variations in apparent size of the Sun and of the Moon and in apparent brightness of the other planets. However Ptolemy, except perhaps in very broad terms, appears to have excluded these implications from consideration. He was concentrating on accounting for directions.

The Copernican revolution

Sometimes Copernicus is given credit not merely for what he himself directly contributed but also for the main lines on which his successors developed his ideas. His contribution was truly great but it is wrong not to recognize the limitations remaining in his theory. For that matter it is wrong to take too literally what the writers of general historical texts say about his theory. For instance, his theory was not strictly heliocentric (but then Ptolemy's was not strictly geocentric). Copernicus' theory was heliostatic in contrast to Ptolemy's geostatic theory. However, the most we would now say is that the Sun is stationary in the context of the planetary system; it is certainly not stationary in the context of the Galaxy. More important is the fact that Copernicus did not remove, as is often suggested, the complexities of deferents and epicycles from his theory. It was Kepler who removed them. It is true that Copernicus dispensed with them as the means of accounting for the alternation of direct and retrograde apparent planetary motions. However, in order to cope with other problems his resort to them was even greater than Ptolemy's. He objected to Ptolemy's use of the equant because it was inconsistent with any literal notion of uniform velocity and substituted for it more epicycles. His objection here reveals how traditional his thinking was.

It is strange that the twin notions of the circular motion and the uniform velocity of celestial bodies should have been sacrosanct from classical Greek times to the seventeenth century AD. They were supported by the belief that while terrestrial motions were imperfect, celestial motions were perfect. Copernicus did a great deal to break down such distinctions between the terrestrial and the celestial but he never doubted the need to analyse apparent celestial motions into basic uniform circular motions. There is, of course, no greater absolute perfection in uniform than in variable velocities or in the circle than in any other figure. There may be simplicity but that is another matter. Furthermore there are no grounds for believing that celestial motions must be either simple or perfect. There is a tendency to equate simplicity or uniformity with regularity and that in turn with lawfulness. However, as we shall see, lawfulness does not require patent regularity and for that matter regularity is not confined to the simple or uniform. There is a further tendency to regard any regularity in the sense of order and pattern as evidence for some superior guidance of events. It is often thought that things, if left to themselves, would be chaotic, and the absence of chaos reveals that there has been planned intervention and that one has got near to the *raison d'être* of the situation when one finds any kind of order. All of this, of course, is mystical thinking.

Though his general approach was traditional, there were revolutionary germs in Copernicus' theory. Amongst these was his placing of the Sun instead of the Earth in the focal region of planetary motion and his recognition that the Earth rotated around its axis in 24 hours and revolved around the Sun in 365¼ days. Jointly these enabled him to disentangle from the complex of apparent celestial motions three components (a) diurnal motion resulting from the Earth's rotation, (b) the annual motion of the Earth around the Sun, and (c) the (sidereal) motion of the Moon around the Earth and the (sidereal) motion of the other planets around the Sun. For the apparent motion of the fixed stars, only the first was required; for the apparent motion of the Sun, the first and second; and for the Moon and the five remaining planets known to the Ancients, all three. In doing so, he recognized the Earth as one of the planets, he took this role from the Sun and he gave the Moon a special role which made it the first recognized satellite. Further he introduced, in a way his predecessors did not, the idea of distance.

Had Copernicus' book been more readily intelligible to non-technical readers, the violent reaction to heliostatic notions would probably not have been delayed for over half a century. Galileo's more general statement, aided no doubt by his more ebullient personality, provoked the storm. Because his case became famous, the Roman Catholic Church is sometimes thought to be the arch opponent of scientific progress at that time. However, the utterances of some contemporary Protestant leaders show them to be equally intolerant. Scientific and theological issues were not in those days as distinct as we now see them to be, or perhaps make them. Servetus, who taught, in the midst of a good deal of theological heresy (as judged by both sides), the pulmonary circulation of the blood, was burned by the Calvinists in 1553. Bruno, who taught, in the midst of a great deal of theological heresy, the Copernican theory, was burned by the Roman Catholics in 1600. There can be little wonder that Galileo with the memory of this burnt flesh in his nostrils was so ready in 1634 to admit his error when the pressure was put on him. He not only had the temerity to support Copernicus but had dared to say what the Scriptures could have meant when they said that the Lord, to aid Joshua, had made the Sun stand still. The Scriptural support for the movement of the Sun seemed as relevant to his contemporaries as any physical observations. Indeed it was more hallowed by time than anything his telescope could reveal.

Copernicus' first main contribution was to attribute the apparent diurnal motion of the celestial bodies to a diurnal rotation of the Earth around an axis passing through the celestial poles. He argued forcefully against the possible objections to the idea of the rotation of the Earth and equally forcefully against the idea that the remote stars could make a complete 360-degree circuit in just 24 hours. The idea of a terrestrial diurnal rotation was not incompatible with the Ptolemaic system, the principal features of which dealt with the motions of the 'planets' relative to the 'fixed' stars and not with the apparent diurnal motions upon which these relative motions were imposed. However, the Ptolemaic notion that the Earth was at rest extended unnecessarily to a denial of its own rotation.

Copernicus' second contribution was in effect a turning inside out of the main features of the Ptolemaic system. By placing the Sun at rest, the Earth in motion around it and the other planets, except the Moon, in motion around the Sun instead of the Earth, he extracted the effect, as we now understand it, of the Earth's revolution upon the apparent motions of the other bodies. The possibility of this could have been more readily seen had the periods of the Ptolemaic epicycles been reckoned in the modern way. The relevant values are exhibited in Table 3.1. It can be seen that the epicycles of Mercury and of

Period of	Mercury	Venus	Mars	Jupiter	Saturn
Deferent	1 yr.	1 yr.	1 yr. 322d.	11 yr. 314d.	29 yr. 167d.
Epicycle (A)	116d.	1 yr. 219d.	2 yr. 50d.	1 yr. 34d.	1 yr. 13d.
Epicycle (M)	88d.	225d.	1 yr.	1 yr.	1 yr.

Table 3.1   Periods of Ptolemaic deferents and epicycles, the latter being reckoned in both the Ancient (A) and the modern (M) way.

Venus and the deferents of the other three have the sidereal periods of these planets, whereas the deferents of the first two and the epicycles of the other three have the period of the Sun (or of the Earth when the latter is thought to be in motion). Thus what Copernicus attributes once to the Earth's motion is needlessly attributed by Ptolemy five times to the planets and a further time to the Sun. An analogous situation occurs with the Ptolemaic ratios of the radii of epicycle and deferent for each planet and the Copernican ratios of the planets' radii to the radius of the Earth, or in more recent terms the mean distances from the Sun—the mean distance of the Earth from the Sun is the astronomical unit of distance, AU. This may be seen in Table 3.2. The third row of this table gives the ratios of the radii of the orbits of the five planets to the radius of the Earth's orbit, according to Copernicus. It is clear that these are equivalent to the ratios of the radius of epicycle to radius of deferent in the cases of Mercury and Venus, and to the ratios of the radius of deferent to the radius of epicycle in the cases of the other three. That is, once again the deferents of the first two and the epicycles of the last three are shown to be the orbit of the Earth around the Sun.

	Mercury	Venus	Mars	Jupiter	Saturn
Ptolemaic ratios	0.3708	0.7194	0.6583	0.1917	0.1083
reciprocals	—	—	1.5215	5.2165	9.2336
Planetary orbit in AU					
Copernicus	0.3763	0.7193	1.5198	5.2192	9.1743
Modern value	0.3871	0.7233	1.5237	5.2028	9.5388

Table 3.2   Ratios of radius of epicycle to radius of deferent according to Ptolemy, and the reciprocals of these ratios for Mars, Jupiter, and Saturn, together with the radii of the planetary orbits in terms of astronomical units (AU) according to Copernicus and to modern findings.

Not only does this disentanglement result in a simplification of assumption but also it avoids the need for an arbitrary assumption that two deferents and three epicycles have the same period as the Sun. Such a coincidence calls for explanation over and above that called for by the brute fact that the Earth requires $365\frac{1}{4}$ days to complete a revolution about the Sun. Though this point cannot be made about the ratios of radii in the two theories, another point can. Copernicus' ratios are, in themselves, as arbitrary or *ad hoc* as Ptolemy's. They were chosen to give the right results. If the ratios are thus, the positions of the planets in the sky must be as we find them to be. In Ptolemy's theory there was no way of checking these ratios apart from establishing that they enabled the correct deductions. Should they prove not to give precise enough results, they can be adjusted so as to do better. In Copernicus' theory, however, it became possible to check his ratios in another way as soon as it became possible to assess the relative distances of the planets from the Sun. Thus, though *ad hoc* in origin, these values were in the context of a heliofocal theory independently testable in a way that was not possible in a geofocal theory.

Before we go further it may be important to show first how Copernicus explained so simply the main features of apparent planetary motion and second how fussily complex his theory had to remain in order to account for detail. The first aim can be achieved by considering the account of Mars' apparent motion in terms of (i) the sidereal periods of Mars and of the Earth, that is, the periods of Mars' and of the Earth's revolutions around the Sun and (ii) the relative distances of Mars and of the Earth from the Sun. The relevant values are set out in Table 3.3.

	Sidereal period (days)	Mean distance from Sun (AU)
Earth	365¼	1.00
Mars	687	1.52

Table 3.3   Periods and mean distances from the Sun of the Earth and Mars.

Omitting refinements we may regard the two bodies as moving with uniform velocity on circular orbits. Let us begin at a time 90 days before Mars comes into opposition with the Sun. The Earth will have almost a quarter of its orbit (90/365¼th) and Mars a little over two fifteenths of its orbit (90/687th) to traverse before Mars is in opposition. These starting points are marked 0 on the two orbits in Fig. 3.12.

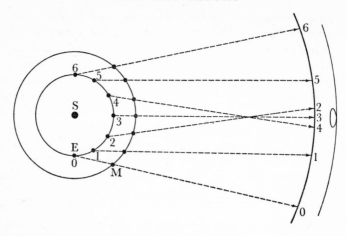

Fig. 3.12 Copernican explanation of the apparent celestial path of an external planet, illustrated with Mars.

Mars will be seen from the Earth at point 0 on the apparent celestial sphere. Thirty days later the apparent position of Mars will be at point 1, and 60 days later at point 2. That is, Mars appears to be moving eastward through the Zodiac with a decreasing velocity. During the next 30 days when the Earth draws level with Mars, the latter appears to move back (westward) in the sky to point 3 on the apparent celestial sphere. This apparent retrograde motion continues for another month. Thereafter as the Earth draws markedly ahead of Mars, the latter resumes its apparent eastward motion with increasing velocity. As in the Ptolemaic theory the minimum distance between the two bodies occurs when Mars is in opposition, accounting for the maximum brightness of Mars at that time, though the Copernican theory insisting on precise relative distances gives a more precise explanation of the varying apparent brightness of Mars. An angular difference between the planes of the two orbits would produce a loop in the apparent path of Mars such as revealed in Fig. 3.3.

Because Copernicus clung to the dual assumptions of circular and of uniform motion, he preserved the eccentrics and the deferent-epicycle systems introduced by Hipparchos and exploited by Ptolemy. He could not put the Sun at the centre of any of the planetary deferents; indeed he could not make the centres of the several deferents coincide, as Fig. 3.13(a) indicates. This left an untidiness that suggests

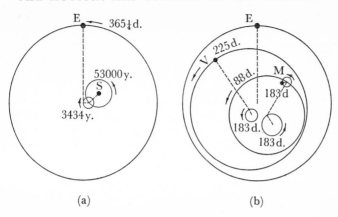

(a)   (b)

Fig. 3.13   Copernicus' analyses of the motions of the Earth (E) in relation to the Sun (S) and of Venus (V) and Mercury (M). None of them literally centres on the Sun; nor do the centres of their motions coincide. (Adapted from A. R. Hall, *The Scientific Revolution*, Longmans, Green and Co. Ltd, by permission of the publishers.)

(no stronger word is appropriate) that something was amiss with the theory.

Some seventy years later, Kepler located what was wrong in the Copernican analysis. Though showing other addictions to mystical thinking, Kepler abandoned uniform circular motion. He showed, using Brahe's carefully gathered data for Mars, that the facts of apparent planetary motion could be accounted for by assuming that (i) the planetary orbits were ellipses with the Sun at one of the foci, (ii) the planetary velocities varied so that radius vectors included equal areas in equal times, and (iii) the square of the period of revolution of a planet is proportional to the cube of its mean distance from the Sun. These three hypotheses have come to be known as Kepler's laws of planetary motion. The first may be illustrated in an exaggerated diagram as in Fig. 3.14(b). If one places two pins in a card and loops a piece of string around them to guide a pencil in tracing out a closed figure, one produces an ellipse. The positions of the pins are the foci of the ellipse. The wider the separation of the foci the more elongated (eccentric) the ellipse. When the two foci coincide the ellipse collapses into a circle, which is thus a special or limiting case of an ellipse. The orbit of the Earth is only slightly eccentric, but its eccentricity is

47

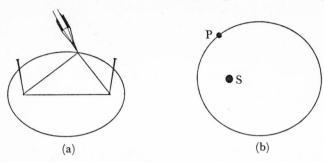

Fig. 3.14 (a) Drawing an ellipse of marked eccentricity. (b) A less eccentric ellipse illustrating Kepler's first law.

sufficient to account, in terms of the second law, for the anomalous change of apparent pace of the Sun through the fixed stars during the course of the year.

Kepler was convinced that the figure of the planetary orbits was elliptical through his analysis of data for Mars which has a more eccentric orbit than the Earth. His analysis of these data also showed him that Mars had a greater real angular velocity when nearest the Sun (i.e. at perihelion) than when furthest from it (i.e. at aphelion). More precisely he showed that when the two 'triangles' A and B (as shown in Fig. 3.15) were equal in area, the times to traverse the arcs on the orbit forming their 'bases' were equal.

Kepler's third law can be appreciated through a consideration of the data presented in Table 3.4. The more distant a planet from the Sun, the greater will the circumference of its orbit be. The circumference of a circle is 2π times its radius, where π is approximately 3.1416. Hence when the mean distance, D_1, of one planet is greater than the mean distance, D_2, of another, the circumference of the

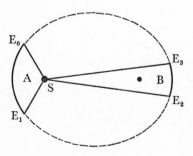

Fig. 3.15 Kepler's second law. If triangles A and B are equal in areas the times to traverse the arcs E_0E_1 and E_2E_3 will be equal. NOTE the similarity with Ptolemy's notion of the equant.

	Mercury	Venus	Earth	Mars	Jupiter	Saturn
Period in years	0.2408	0.6152	1.000	1.881	11.862	29.46
Mean distance in AU	0.387	0.723	1.000	1.524	5.203	9.539
Period squared*	0.058	0.378	1.00	3.54	140.7	867.9
Mean distance cubed*	0.058	0.378	1.00	3.54	140.7	867.9

*These values are rounded.

Table 3.4   Illustrating Kepler's third law, namely that the square of the planet's period is proportional to the cube of its mean distance ($P^2 = D^3$ where the Earth's period and mean distance are set to one).

former will be greater than that of the other in the ratio of approximately D_1/D_2 (assuming the orbits are approximately circular). Thus, as Jupiter has a mean distance of 5.203 AU from the Sun and Venus a mean distance of 0.723 AU, the circumference of Jupiter's orbit is approximately 5.203/0.723, that is, approximately 7.2 times greater than the circumference of Venus' orbit. If the two planets moved with equal velocity around their orbits, Jupiter's period would be approximately 7.2 times that of Venus'. However 7.2 times 0.6152 years (Venus' period) is approximately 4.43 years, whereas Jupiter's period is 11.862 years. That is, Jupiter's angular velocity is distinctly less than Venus'. Kepler's third law asserts that the more distant the planet the less its angular velocity. This relationship is analogous to that stated in the second law, namely the angular velocity of a planet is greatest when it is nearest the Sun. Kepler had found here a single way of stating what Ptolemy had to state separately for each planet as pairs of *ad hoc* values for deferent and epicycle, and what Copernicus had to state as connected pairs of *ad hoc* values for the Earth and each planet in turn.

For the benefit of those who believe that laws of nature are discovered by generalizing from large numbers of cases, it is worth stressing that Kepler's three laws of planetary motion were 'discovered' by his persistent trial of many hypotheses on the data for a single planet, Mars. It is also worth mentioning, to illustrate how irrational the source of an hypothesis may be, that at one stage Kepler tried to fit his data to an oval (egg-shaped) orbit in the belief that as the egg is

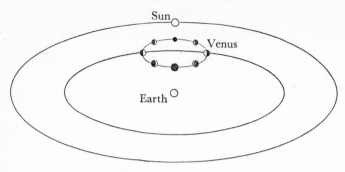

Ptolemaic implication

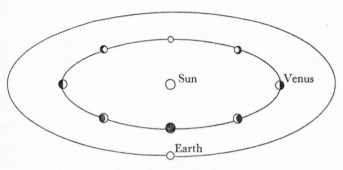

Copernican implication

Fig. 3.16   Phases of Venus as deduced from the Ptolemaic and the Copernican theories. In both cases it is assumed that Venus is seen in light reflected by it from the Sun. According to the Ptolemaic theory the half of the sphere illuminated by the Sun would never be turned directly towards the Earth and hence there would never be a 'full' Venus. Galileo saw that there was.

the beginning of life its shape might be the figure of real planetary motion.

While Kepler was putting Copernican theory into a neater, and more modern, form, Galileo was producing (i) new evidence compatible with that theory but not with the Ptolemaic and (ii) dynamic notions which when elaborated by Newton transformed the Copernican theory from a mathematical to a physical theory. Though an important figure in the history of theory, Galileo was in many ways more important as an observer and an experimenter. He was not, however, in any sense a mere fact-grubber. He recognized the theoretical importance of what he observed; indeed, sometimes he made the observation only because he recognized the theoretical significance of the fact it could yield.

His new evidence for planetary theory came mainly from his development for astronomical purposes of the telescope which had been invented by the Dutch. Many of his new facts were circumstantial rather than crucial in deciding between the Ptolemaic and Copernican theories. Sunspots seen for the first time showed the Sun to be rotating and to be imperfect (irrelevant to the main issue but quite telling against the traditional distinction between the imperfect terrestrial and the perfect celestial). The revelation that the visible markings on the Moon were produced by mountains whose heights could be estimated from their shadows served to relate that celestial body to the Earth. The revelation that each of the star-like planets had a visible disc (if we except the peculiar shape of Saturn in Galileo's telescope), like the Sun and the Moon, also served to tie the members of the Solar System together. More importantly the telescope revealed the phases of Venus, running like the Moon from crescent, through half and gibbous, to full. This is quite inconsistent with Ptolemaic theory if one grants that the appearance of Venus is the result of its reflection of light from the Sun. In that event it would run from new (or thin crescent) to half to new, back to half and then to new again in the course of its synodic period. Galileo saw that it ran from thin crescent (shortly after inferior conjunction), through half (at quadrature), to full (at superior conjunction when not obscured by the Sun), through half, back to thin crescent. Though this was implied by the Copernican theory, it was quite inconsistent with the Ptolemaic (as Fig. 3.16 shows). Perhaps more disturbing to Ptolemaic theory was Galileo's observation of the four major satellites of Jupiter, the Medicean planets as he called them in order to flatter his patron, who was a member of the de Medici family. Their apparent motions were hardly consistent with any other proposition than that they revolved around Jupiter. If they did indeed revolve around Jupiter rather than around the Earth, why should not other planets revolve around the Sun rather than around the Earth? A good question, but not one forcing any one answer. However, the evidence that made it a rational question to ask (when tradition suggested it to be irrational) also suggested the answer we now regard as proper.

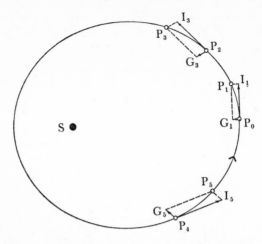

Fig. 3.17   At aphelion P_0 the instantaneous velocity of the planet is perpendicular to the radius vector from the Sun. Were the planet moving under inertia alone it would reach point I_1 in a given time; were it moving under gravitation alone it would reach point G_1 in that time; the joint forces bring it to point P_1. Gravity has pulled it inwards and has speeded it up. This speeding up is accentuated at points like P_2 where the directions of inertial motion and gravitational motion form an acute angle. Such speeding up occurs between aphelion and perihelion. At points like P_4 where the directions of the two motions form an obtuse angle, gravitation slows the planet down because the pull is against rather than with the inertial motion.

It would take too great a space to state Galileo's other contribution, namely his early formulation of mechanics. It includes his discovery of the way free-falling bodies accelerate and of the independence of their period of fall (that is, setting aside the resistance to their fall by the medium such as air and water through which they fall) of their weight. Galileo gave Newton a base from which he could transform Kepler's remodelled Copernican system. Newton used two main dynamic principles—the law of inertia and the law of universal gravitation. Prior to Galileo, it had been assumed not only that a body at rest would remain so unless some force was applied to it but also that a body in motion would come to rest unless some external force continued to be exercised on it (or until the momentum imparted by some external force was exhausted). Newton's law of inertia asserts that a body in motion will continue to move with uniform velocity in a straight line unless affected (speeded up, slowed down or diverted) by some external force. Thus, a planet in orbit around the Sun would, in accordance with the law of inertia, fly off at a tangent as a stone released from a sling, unless some force continued to curve it in towards the Sun. The action of the latter force is specified in the law of gravitation which asserts that particles of matter attract each other in direct proportion to the product of their masses and in inverse proportion to the square of their distances, that is:

$$F = G\, m_1 m_2\, /\, r^2$$

where G is the gravitational constant, m_1 and m_2 are the masses of the particles concerned, and r is the distance between the particles. Thus the greater the distance of a planet from the Sun the less the mutual gravitational attraction.

Though rigorous mathematical argument alone can show that Kepler's laws are implied by Newton's laws when taken in conjunction with certain facts, the following comments may give non-physicists some intuitive understanding of the relationship. Consider a body P moving, as represented in Fig. 3.17, from point P_0 in the direction of point I_1 with an instantaneous velocity of a magnitude such that in a brief interval of time it would, if not subject to some external force, move in accordance with the law of inertia, to I_1. However, let it in fact be subject to a gravitational attraction towards S such that it would move to G_1 were it not in inertial motion. The resultant displacement, in accordance with the parallelogram rule, is from P_0 to P_1. As the two 'tendencies' are interacting continuously, the path of P is a curve and not a succession of short straight lines as the parallelogram rule suggests. Thus the resultant displacement of P is on a path curving in towards S. Newton showed that the curve must be one of the family of conic sections, S being at a focus (*see* Fig. 3.18). Where P is a planet in orbit around the Sun, the curve being closed is an ellipse. A circular orbit is a special case where the foci collapse to a single point. Where the curve is not closed, as may be the case with some comets, it will be a parabola or an hyperbola.

Kepler's second law can be understood through reference to the three constructions in Fig. 3.17. It can be seen that as the planet moves from P_4 to P_5 on its way to aphelion it is slowed down by gravitation as well as curved in towards S. After it passes aphelion

53

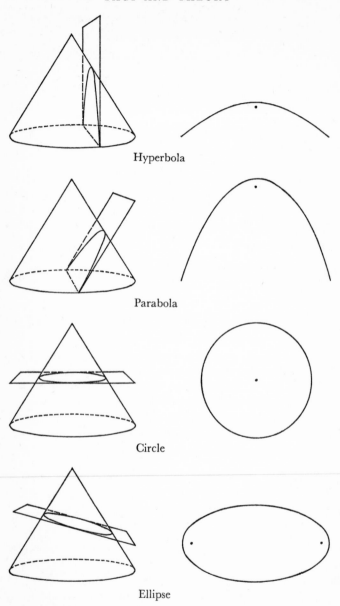

Hyperbola

Parabola

Circle

Ellipse

Fig. 3.18   Conic sections. The diagrams on the left show the plane of the section and the diagrams on the right the resultant figure.

at P_0 it is speeded up by gravitation. At aphelion and perihelion the instantaneous velocity of P is perpendicular to the radius vector SP. Elsewhere gravitation which is in the direction of S pulls P back as at P_4 or pulls it on as at P_2. Thus on the journey from perihelion to aphelion the planet is slowed down and from aphelion to perihelion speeded up.

Finally, it can be appreciated that the distance and the period of the planet must vary together. As the distance of P from S increases, the gravitational force exerted on P decreases (as the square of the distance). If P is not to escape, its mean instantaneous velocity must decrease as the mean distance from S increases. That is, as the diameter of the orbit increases, the speed of the planet decreases; thus having a longer journey at a slower speed, its period must increase at a greater rate than its distance from the Sun. Kepler's third law states this in a more precise form which is deducible from Newton's principles.

Newton also showed that his laws implied a number of other phenomena. For instance, the Earth having a slight bulge at the Equator and an axis inclined to the plane of the Ecliptic, the force of gravitation will lead to a slow wobble in its axis of rotation. This is manifested in the precession of the equinoxes. A similar effect is found in the retrogression of the nodes of the Moon. Gravitation also accounts for the tides. Galileo had seen that the tides had some relevance for the heliocentric theory, but even though he seems to have cheated with the data he never succeeded in showing this relevance.

Newton's theory of mechanics is regarded as explaining all the mentioned phenomena because they may be deduced from it. They follow from it in the same general way that the theorems in geometry follow from the axioms, the definitions and the 'givens', granted the rules of inference.

A qualification needs to be added to the foregoing account of the way Newton's laws explain Kepler's laws. It has already been said that this explanation is given when Newton's laws are taken in conjunction with certain facts (e.g. the masses and distances of the Sun and the planet and the instantaneous velocity of the planet at a given point). What must be added is that the implied elliptical orbit and changes of angular velocity as specified in Kepler's first and second laws follow only in a two body (the Sun and one planet) situation. Two planets, according to Newton, will have some appreciable gravitational effect on each other, although departures from Kepler's laws would be slight where the masses of the planets are small relative to that of the Sun and/or their distance from each is great. Observation shows that Kepler's laws are at best good approximations and that what the Newtonian laws imply is a much better approximation.

The spacing of the planets

Though from very early times the planets were placed in an order of distance from the Earth, no relative distances were demanded until Copernicus propounded his theory. Ptolemy adopted the following order of distance from the Earth: the Moon, Mercury, Venus, the Sun, Mars, Jupiter and Saturn. Copernicus naturally thought of distances from the Sun. The values, in astronomical units, that he proposed in order to make his theory work, were Mercury 0.38, Venus 0.72, Earth (with the Moon) 1.00, Mars 1.52, Jupiter 5.22 and Saturn 9.17. This spacing, later established more accurately, engaged the attention of several astronomers in the two centuries after Copernicus. We shall consider the views of just two of them.

Kepler in 1596, before he turned his agile mind to the problem of the figure of planetary orbits, considered the spacing of the planets. He decided that if the planetary orbits were regarded as defining a nest of spheres, the spaces between the spheres would neatly accommodate a series of regular solids. Thus a cube just contained within the sphere defined by Saturn's orbit would just contain the sphere defined by Jupiter's orbit. If the distance from the centre point of a cube to the nearest point on a face is taken as 1.00, the distance from the centre to a corner is 1.73 approximately. Thus a sphere just containing a cube will have a radius of 1.73 times greater than the radius of a sphere just contained by that cube. The accepted radius of Saturn's orbit (9.17 AU) was 1.76 times greater than the accepted radius of Jupiter's orbit (5.22 AU). Kepler considered that 1.73 (his theoretical value) and 1.76 (the then accepted factual value) were remarkably close, allowing for the great distance of Saturn from the Sun and the consequent difficulties in estimating that distance precisely. In similar manner he found that a tetrahedron (a solid with four triangular faces) would just fit within the spheres defined by the orbits of Jupiter and of Mars, a dodecahedron (a solid with twelve octagonal faces) within the spheres defined by Mars and of the Earth, an icosahedron (a solid with twenty triangular faces) within the spheres

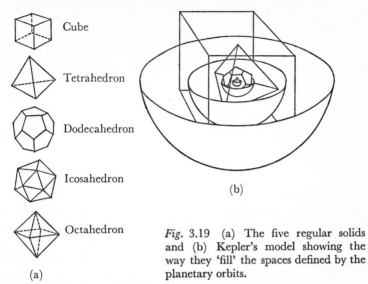

Cube

Tetrahedron

Dodecahedron

Icosahedron

(b)

Octahedron

Fig. 3.19 (a) The five regular solids and (b) Kepler's model showing the way they 'fill' the spaces defined by the planetary orbits.

(a)

defined by the Earth and Venus and an octahedron (a solid with eight triangular faces) within the spheres defined by Venus and Mercury. This piece of 'solid-fitting' (*see* Fig. 3.19) was remarkable in its closeness. But even more remarkable was that there were just five spaces to be filled and that the five solids used to fill them constituted the total supply of regular solids. It seemed implausible that this could be merely a coincidence. Its aptness, the complete meshing of differently derived properties, gave it for Kepler some mystic, some cryptic significance. The Pythagorean philosophers of Ancient Greece looked for numerical and geometrical relations which seemed to give such cryptic hints. That $3^2 + 4^2 = 5^2$ seemed to them to be such a case. The geometrical theorem that the sum of the squares on adjacent sides is equal to the square on the hypotenuse of a right-angled triangle, a more general statement of the arithmetical relationship just given, is still known as the theorem of Pythagoras. There seemed to be some quality of revelation in this. What comes out so neatly in an unexpected way—an improbable aptness, as it were—carries some suggestion of revelation of deep-lying truth. There is an affinity, in the judgement of some, between these 'revelations' and self-evident, necessary truths. Modern science, possibly excluding mathematics, is no longer engaged in the conscious pursuit of necessary truths, a

57

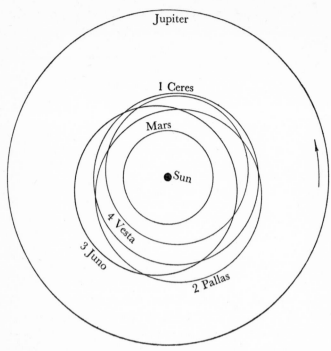

Fig. 3.20 The orbits of the first four discovered planetoids.
(Adapted from *The Flammarion Book of Astronomy* and reproduced
by permission of the publishers, Allen & Unwin Ltd, London.)

pursuit which bedevilled the earlier history of science. Nevertheless
this same devil may now and then be found riding on the shoulders
of some branch of contemporary empirical science.

There was an element of luck or ill-luck, however one cares to view
it, in Kepler's *tour de force*. First, the modern values of planetary spacing
show the degree of fit to be much poorer than Kepler imagined. For
example, the ratio of Saturn's to Jupiter's orbit is really about 1.83,
which is a fair way from the required 1.73. Second, it turns out that
there are more than six planets in the Solar System and hence more
spaces to fill than there are available regular solids. The mystical
aptness has thus vanished.

In 1772 Bode, following a lead given by Titius, noted that the mean
distances of the planets from the Sun expressed in astronomical units
could be closely approximated in the following numerical operations:

Write the series of numbers $4 + (0 \times 3)$; $4 + (1 \times 3)$; $4 + (2 \times 3)$; $4 + (4 \times 3)$; $4 + (8 \times 3)$; etc. and divide each by 10 to obtain:

0.4	0.7	1.0	1.6	2.8	5.2	10.0

If a gap is left between Mars and Jupiter, the mean distances from the Sun as we now know them approximate these values, thus:

0.39	0.72	1.00	1.52	—	5.20	9.54
(Mercury)	(Venus)	(Earth)	(Mars)		(Jupiter)	(Saturn)

In 1781 Herschel discovered the planet Uranus with a mean distance from the Sun of 19.2 AU. By extrapolation, Bode's law required that a planet beyond Saturn should there be one, to be at $(4 + 192)/10 = 19.6$ AU which is fairly close to the fact. Such near corroboration of the law naturally led to speculation about a planet occupying the 'gap' between Mars and Jupiter. At the end of the eighteenth century co-operative plans were made to look for the 'missing' planet. However, before they could be put in effect, the planetoid Ceres was observed (1801). And soon after Pallas, Juno and Vesta were observed. These four planetoids have orbits with much the same semi-diameters, approximately the 2.8 expected from Bode's law (*see* Fig. 3.20). Thus the law, despite an arbitrariness which we shall discuss in a moment, seemed to point to a regularity from which prediction was possible even if a rationale of planetary spacing was not available.

Quite apart from the arbitrariness of a series involving multiples of 3 (rather than any other number), there is a particular arbitrariness in Bode's series. The value preceding (1×3) in a series (1×3), (2×3), (4×3), (8×3) should be (0.5×3) and not (0×3). The law may be made to look more respectable by stating it in the form:
$$d = a + bc^n$$
where $a = 0.4$, $b = 0.3$, $c = 2$, and $n = -\infty$ for Mercury, 0 for Venus, 1 for the Earth, 2 for Mars and so on. However, this merely hides the arbitrary values of 0.4 and 0.3 and does little to hide the peculiarity of the series of indices $-\infty$, 0, 1, 2, etc.

The discovery of Neptune in the late nineteenth century and Pluto in the early twentieth century showed that the law as stated could not be sustained. The mean distances projected from the law for the next two planets beyond Uranus are:

$$4 + (128 \times 3)/10 = 38.8 \text{ and } 4 + (256 \times 3)/10 = 77.2.$$

59

The mean distance of Neptune from the Sun is 30.07 AU and of Pluto 39.52 AU. If Neptune could be rejected as a legitimate member of the family of planets, then Pluto's 39.52 might be passed off as a reasonable approximation of the 38.8 to be expected for the next planet beyond Uranus. However, if either Neptune or Pluto is a changeling, Pluto is the more likely candidate—it could be an escaped satellite of Neptune.

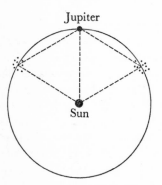

Fig. 3.21 The two Lagrangian points form equilateral triangles with the Sun and Jupiter. Members of each of the two sets of Trojan planetoids oscillate around one of these points.

Though Bode's law must be regarded as a mistaken identification of a regularity, there is a strong suggestion from other considerations that some conceivable spacings of the planets are possible and others not. In terms of Newtonian theory, it can be shown that with two or more bodies in orbit around a third, the mutual gravitational fields render some positions stable and others unstable. Lagrange (1736-1813) showed that if two bodies had the same orbit around a third, the only stable positions for them would be at the apices of an equilateral triangle with the third body. There are a number of planetoids, the Trojan planetoids, with roughly the same orbit as Jupiter (*see* Fig. 3.21). Several of them travel ahead of Jupiter and several behind. One set oscillates around a point forming the third apex of an equilateral triangle with Jupiter and the Sun, and the other set oscillates around a point forming the third apex of the other possible equilateral triangle with Jupiter and the Sun. A second illustration of the notion of stable and unstable positions in a complex system of bodies in orbit is provided by the gaps in the sequence of orbits of the many thousands of planetoids discovered since 1801. There are no planetoids having orbits with periods of revolution which are simple fractions of the period of

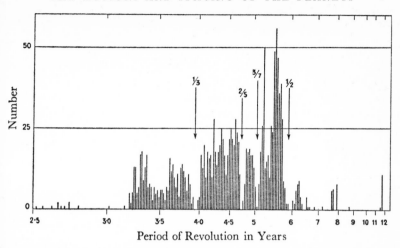

Fig. 3.22   Frequency distribution of planetoids of various periods
of revolution. NOTE the gaps at $\frac{1}{3}$, $\frac{2}{5}$, $\frac{3}{7}$, $\frac{1}{2}$, etc. of Jupiter's period
of revolution. (Adapted from *The Flammarion Book of Astronomy* and
reproduced by permission of the publishers Allen & Unwin Ltd,
London.)

Jupiter's revolution such as $\frac{1}{3}$, $\frac{2}{5}$, $\frac{3}{7}$ and $\frac{1}{2}$ (*see* Fig. 3.22). The attraction
by massive Jupiter of planetoids originally in these orbits would act
repeatedly in the same way. Hence their orbits would be substantially
changed over a long time.

Thus it would seem that although the planets need not, as they do
not, take up positions in strict accordance with Bode's law, they
could not maintain stable positions in certain orbits. In so far as
Bode's law hints at this it is sound, but in detail it is clearly wrong.

Some tests of the heliofocal theory

It has been pointed out that the theories of planetary motion propounded by Eudoxos and by Ptolemy were effective, within given
limits, partly because values had been chosen for the axes, periods
and direction of rotation of spheres or for the periods, radii and
inclination of deferents and epicycles which would give the required
results. If the original data of observation later proved to be inaccurate,
then some of these values could be changed to give the new results.

Sometimes, of course, new spheres, or new epicycles would be required. Copernicus' theory, especially as modified by Kepler, introduced a new feature. In order to lead to the right results in respect of apparent motions in the sky (the phenomena to be accounted for), the heliofocal theory required the planets to be spaced in a certain way. Once it became possible, through the improvement of devices for measuring the direction of members of the Solar System from different points on the Earth's surface, to assess the distances between the members of that system, an independent test of the otherwise *ad hoc* values was provided. No such independent test existed for the theories of Eudoxos and of Ptolemy.

A second independent test of the heliofocal theory was also early recognized. As we now know the Earth has a mean distance of some 93,000,000 miles from the Sun. Thus over a six month period the base line for the observation of any fixed star extends over 186,000,000 miles. If this base line is not negligible in relation to the distance of some fixed star, the apparent direction of that star will vary with the Earth's position on the base line. Unfortunately early attempts to establish such shifts of apparent direction were unsuccessful. The then accepted distances of the fixed stars were gross underestimates and the technical means of detecting the very small parallax shifts, which in fact exist, were not available. The story of the search for the parallax shifts is worth telling here if only for an important by-product, an important piece of serendipity.

Though Eudoxos may have thought that the fixed stars were equidistant and attached to the one celestial sphere, modern belief has always taken them to be at different distances. If this were not accepted before Galileo, his telescopic observation of an incomparably greater number of stars than were seen with the naked eye strongly suggested that the fixed stars were not at the one distance. If this were so, and if the Earth changed its position, then the nearer of the stars should appear to shift in a six-month period relative to the more distant stars. No such shifts, however, had been detected up to the early part of the nineteenth century. Tycho Brahe, in the late sixteenth century, estimated that the nearest stars might be no more distant than 6,000 times the distance of the Earth from the Sun. He worked out that if the Earth were in orbit around the Sun, the nearest stars would have a parallactic displacement of almost one minute of arc. His technical means of measuring angles might well have enabled him to detect such a displacement. As he could find no such displacement, he argued against the Copernican theory. His estimate of stellar distances, however, was seriously in error. Sirius and Procyon, the two nearest very bright stars visible from European latitudes, are approximately 557,000 AU and 723,000 AU distant respectively. Rigel, a star a little brighter than Procyon, is about 55,788,000 AU distant. Thus Brahe's estimate of the smaller stellar distances was only a ninetieth of the distance of the nearest very bright star visible to him, and less than a 9,000th of the distance of some other very bright stars. This put the observation of any parallax quite beyond his technical powers of observation.

When Bradley in 1745 looked for a stellar displacement through the parallax effect, he did so, not surprisingly, with much greater sophistication than Brahe. He had the advantages given by over a hundred years' very considerable factual, technical and theoretical advances in astronomy. He chose a star, *gamma Draconis*, which, because it passes near the zenith from his place of observation, would be little affected by atmospheric refraction. He set up a telescope in a fixed position (strapped to a chimney!) and equipped it with an eyepiece enabling accurate measurements of lateral displacement. The position of *gamma Draconis* in the field of fixed stars is such that any apparent displacement through parallax would be on an ellipse, the major axes of which are oriented so that the extreme displacements would occur in December and June. Bradley found a displacement of about 40 seconds of arc, but, the extremes occurred in September and March. He recognized that this could not be parallactic displacement and recognized it to be a phenomenon known now as the aberration of light. When a person walks in rain which is falling straight down, he, himself, seems to be walking into drops falling in towards him. Thus a drop at head height may be some inches in front but by the time it has fallen to knee height he has moved forward the distance the drop had been in front. The same thing would occur with light if the Earth is moving and if light has a finite velocity. Thus Bradley had produced evidence of an unexpected sort that the Earth is moving as well as evidence that light has finite velocity.

When Bessel, Henderson and Struve each observed stellar parallaxes in the years from 1838 to 1840, they made observations of stars much closer than *gamma Draconis*. For instance, Henderson, working at the Cape Observatory in South Africa, made his observations on *alpha*

Centauri, the brighter of the two Pointers to the Southern Cross. He found a parallax of 0.91 seconds of arc. More refined later measurements give the value as 0.758 seconds which shows the star to be about 272,000 AU distant. Bessel observed 61 Cygni, a faint star, with a parallax of 0.296 seconds, which places it at about 698,000 AU.

It was nothing in the heliofocal theory proper which misled the early investigators in their attempts to establish stellar parallaxes. It was their gross underestimation of stellar distances and the technical limitations of their devices. Though the heliofocal theory implies that there are stellar parallaxes relative to the Earth, the extents of those parallaxes are determined by the distances of the stars relative to the diameter of the Earth's orbit. The parallax is inferred from two considerations or premises: (i) the motion of the Earth and (ii) the distance of the star in AU. What is being tested initially in observing a parallax is the first premise. In so far as the second premise is only a supposition, the test is not a rigorous one. It is always possible to reformulate the supposition to suit one's wishes, that is, either to defend or to attack the theory stated in the first premise. The test becomes rigorous when the second premise consists of a well-established observation and when the inferred effect is well within our technical powers of observation.

Another demonstration that the Earth is in motion is provided by the shifts in the spectra of stars. Newton showed that when sunlight is passed through a prism it fans out to give the 'colours of the rainbow'. Sunlight may be thought of as being made up of light of different wave-lengths. The long wave-lengths, which mediate the perception of red, are bent least by the prism, and the short wave-lengths, which mediate the perception of violet-blue, are bent most. The stars have distinctive spectra or patterns of light wave-lengths. By analogy with sound, these patterns should change when the Earth is travelling towards and travelling away from the stars concerned. A whistle of a train approaching an observer seems to rise in pitch, that is, to shift towards shorter wave-lengths and to fall as the train recedes, that is, to shift towards the longer wave-lengths. Thus when the Earth is moving towards a star there should be a shift in the spectrum of the star towards the violet-blue end and when the Earth is moving away there should be a shift towards the red end. Such semi-annual shifts have been found.

A remarkable test of Newton's explanation of the planetary orbits is provided by attempts to explain in terms of his theory what are

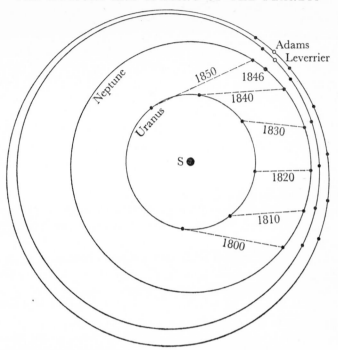

Fig. 3.23   The positions of Neptune as 'predicted' by Adams and by Leverrier in the 1830s and as 'calculated' after its observation by Galle in 1846. (Adapted from *The Flammarion Book of Astronomy* and reproduced by permission of the publishers Allen & Unwin Ltd, London.)

termed perturbations in those orbits. The orbit of a planet would in accordance with the theory be strictly elliptical only if the Sun and the planet were the sole bodies in the system. Other planets pull the planet specified away from its elliptical path in proportion to their mass and to their nearness; indeed they pull the Sun away from the focus of the whole system. When Uranus had been observed for some time, it became clear that there were perturbations in its orbit that could not be accounted for by reference to any other known planet. In the early 1840s Adams and Leverrier working independently of each other showed that these perturbations would be accounted for if there were a hitherto unobserved or unrecognized planet on a given orbit outside that of Uranus, having a given mass and being at a

65

stated point on its orbit at a given time. Neptune, as the extra-Uranian planet has been named, was observed and recognized by Galle in 1846 close to the positions indicated by Adams and by Leverrier (*see* Fig. 3.23). It is true that its orbit, its period and its mass were not quite what had been supposed. Thus Leverrier supposed its mean distance from the Sun to be between 35 and 38 AU (it is 30.07 AU), its period of revolution to be between 207 and 233 years (it is 164.8 years) and that its mass to be 32 times that of the Earth (it is 17.3 times). Adams supposed it to have a slightly larger orbit and period and a distinctly greater mass than Leverrier did: these features are, of course, inter-locked, as the greater the mass of the supposed planet the further it would need to be from Uranus when they were in conjunction (and so from the Sun) and hence the longer its period would have to be. Here again we have a supposition made to save a theory from disproof and the subsequent demonstration by observation that something approximating to the supposition does obtain in reality.

There is a peculiarity in the orbit of Mercury that could be explained in a similar way. Newtonian theory requires that a planet return to perihelion at the same position in space time after time. The perihelion of Mercury was found by Leverrier to be advancing at a rate which could be accounted for in part only by reference to gravitational effects exercised by the known planets. He therefore supposed an intra-Mercurial planet (he named it Vulcan) in order to account for the discrepancy. Lescarbault in 1859 thought he had observed this hypothetical planet when he observed a small round black spot on the Sun. However no other plausible evidence of it has become available. Subsequently Einstein showed that Relativity Theory, granted the Sun is a perfect sphere, can account for the discrepancy between the facts and the gravitational effects of the other planets on Mercury's orbit. As the velocities of bodies increase, Newtonian theory becomes, according to Relativity Theory, a poorer and poorer approximation to reality. The velocity of Mercury is sufficient to show this. Thus the absence of substantial evidence for an intra-Mercurial planet and the existence of an adequate alternative theory led to the recognition that Newtonian gravitational theory is not correct in all circumstances.

4 COMMENTS

VARIETIES OF FACTS AND THEORIES

Alternative theories

Our second case study presented several alternative theories aimed at accounting for the same set of primary phenomena, the apparent motions of the wanderers in the sky. Galen's and Harvey's theories were not alternative in this sense. They both set out to say how the blood moved in the blood vessels but in doing so they assumed the phenomena and not merely the rationale of them to be different. On the other hand, if we overlook for the moment the increasing demand for precision over the centuries, Eudoxos, Ptolemy, Copernicus, Kepler, and Newton propounded theories meant to account for the same basic phenomena. As I shall remark later, they provide illustrations of three varieties of theory. My present concern, however, is with the possibility of alternative ways of accounting for a set of observed facts. With some qualifications to be made later, the planetary theories explain the facts of apparent planetary motion by implying them. That is, if circumstances were as assumed in the theory, the facts to be explained would follow as a logical necessity. Later we shall need to examine carefully the notion 'following as a logical necessity', but for the moment it will suffice to recognize that the theory provides, at least in part, the premises from which the facts may be inferred as a conclusion. That is, we cannot consistently accept the theory (the premises) and reject the facts (the conclusions flowing logically from the premises). For example, if the planet Venus were borne on the inner member of a nest of four spheres rotating in the periods and on axes placed as Eudoxos assumed, then the planet's path in the sky would be as we observe it to be (in its broad features, if not in its precise details). Equally, of course, if Venus moved on an epicycle borne on a deferent, the ratio of the radii and the periods of revolution being as Ptolemy assumed, then the planet's path would also be as we observe it.

It is a point of very great significance that different theories may imply the same observed facts. There arises first the practical problem

67

of deciding which one of the alternatives to adopt. Several criteria of choice have been offered. Other things being equal, the simplest theory, that is, the theory which makes the least, or the least elaborate assumptions, is to be preferred. This criterion is sometimes called the canon of parsimony and sometimes Occam's razor. William of Occam (Ockham in Surrey, England) is credited with the dictum: *Enti praeter necessitatem multiplicandum non sint*, (theoretical) entities should not be multiplied beyond necessity. If the somewhat fussy additional assumptions made by Copernicus in order to account for some of the finer details of apparent planetary motion are ignored, Copernicus propounded a simpler theory than Ptolemy. His principal economy consisted in accounting for the whole of the Sun's apparent motion and a component of the apparent motion of Mercury, Venus, Mars, Jupiter and Saturn by means of the single assumption that the Earth was in motion around the Sun.

A second and more important philosophical significance of alternative theories accounting for the same observed facts resides in the conclusion that a theory which accounts for the facts does not have to be accepted. Though Ptolemy explained some phenomena which Eudoxos did not, and explained some phenomena in more exact terms, it is conceivable that Eudoxos or a disciple could have matched Ptolemy by introducing additional spheres or by setting variable rates of rotation of the spheres or by some other ingenuity. The crucial point is not that of making a decision between the two but that of recognizing that in principle there is potentially no end to the alternative theories which would do equally well. Thus a theory is not necessarily the right one simply because it is the only one which one can think of to explain the facts relevant to it.

Quite often alternative theories which have been propounded differ from one another in ways other than simplicity or economy of assumption. They may differ in respect of the range of the phenomena they explain or in the precision with which they explain the relevant phenomena. Further, one may imply a state of affairs which is contrary to what observation reveals, whereas another implies the state of affairs which is observed. We have seen that, granted that the planet Venus is visible to us because of the light from the Sun reflected by it, Ptolemy's and Copernicus' theories have different implications for the pattern of phases of Venus seen from the Earth. It is not merely that Ptolemy's theory does not imply the observed phases of Venus (as it does not assume or imply any particular spacing of the planets) but that it implies what, in terms of observation, are the wrong phases. Thus if the phases of Venus are not mistakenly observed, Ptolemy must be wrong. This does not automatically make Copernicus right. Tycho Brahe, Kepler's master, proposed a planetary theory which also implied the phases of Venus as observed. He kept the Earth fixed and the Moon and the Sun in revolution around it as Ptolemy did, but he assumed, as Copernicus did, that Mercury, Venus, Mars, Jupiter and Saturn revolved around the Sun. However successful the theory, which was proposed by Copernicus, corrected by Kepler and elucidated by Newton, has proved to be in explaining a wide range of celestial phenomena, there is always the possibility that some other theory will do as well or even better. Indeed in matters of detail Einstein's theory of relativity has improved on Newton's where it has different implications, as in the case of the motion of the perihelion of Mercury.

Types of theory

Several types of theory have been revealed in our first two case studies. At the one extreme are the assumptions by Harvey that the arteries are linked by pores in the flesh with the veins and that the blood consequently is enabled to move in a complete circle around the cardio-vascular system, and the assumption by Adams and by Leverrier of a planet with an orbit beyond that of Uranus. In both cases the theories assume matters of a factual kind, that is, something that can be observed. Harvey's first assumption was converted by Malpighi, and Adams' and Leverrier's assumption by Galle into matters of observed fact. It is not, however, the later transformation of the assumption into a fact which distinguishes this type of theory. Galen's pores in the septum of the heart and Leverrier's intra-Mercurial planet were never reliably observed, yet their assumption must be classed with the other examples given. This type of theory has two distinctive features. First, what is assumed is something which could be observed to be so or not so, were circumstances favourable, and second, what is assumed plays a role like that played by any other important relevant fact. In short, what is assumed serves to round out the factual picture partly revealed by the matters which have been observed.

At the other extreme are theories which do not simply supply

69

missing pieces of the jigsaw puzzle but which provide another sort of rationale for what has been observed. They perform this role by providing bases from which observed or observable situations may be inferred. That is, they imply these situations and so explain them, and predict them if they are not already observed. It is one thing to guess the answer to some problem and another to work it out on the basis of some other considerations whether these be guessed or not. For instance, I may 'judge' from past experience with pieces of sheet-metal of various shapes, that a given quadrilateral figure has an area of 16 square inches; such a judgement has something of the informed guess about it. On the other hand, I may 'judge' the figure to be square and to have sides of 4 inches, and then I may apply to these informed guesses the rule 'the area of a square is equal to the square of its sides' and so infer its area to be 16 square inches. The second type of theory is an assumption of something akin to the rule in this example and not of something akin to the measurable figure and length of sides.

In between these two extremes there is another type of theory even better represented by Mendeleyev's periodic table which figures in the third Case Study (*see* p. 92) than by Bode's law. It is not so much the assumption of any given observable fact or of a general principle of explanatory value as an assumption of the pattern that obtains in a complex set of facts. Its detailed discussion will be left till later.

Amongst the planetary theories we may distinguish three varieties of explanatory theory. Eudoxos provided as the bases from which the apparent celestial motions could be deduced, a series of mechanisms which taken one at a time could be constructed in model form. Eudoxos seems not to have regarded the homocentric spheres as missing parts of the observable jigsaw puzzle. He spoke of the spheres as being invisible and though the Greek word *axon* means both an axle and an axis, it is not likely that he thought of an observable material shaft passing through the Earth and the celestial poles. His rationale for the apparent celestial motions consisted of imaginary mechanisms or mechanical models from which the observed facts may be inferred. Were these imaginary models constructed, they would generate the apparent celestial motions which Eudoxos set out to explain. As he showed in this way that the complex and rather wayward apparent motions of the wanderers in the sky were deducible from uniform circular motions, he gave the former at least a semblance of regularity and lawfulness. A fuller explanation would have been afforded had Eudoxos been able to suggest what drove the imaginary or metaphorical spheres. The Babylonians, who were better as observers and organizers of observed data than as theorists, had a very simple 'explanation' of the wayward motions of the wanderers in the sky. Each had its own angel to propel it and it moved in the path he chose for it. This, of course, affords no real explanation for we have no way of knowing the angel's plan for his planet except by arguing back from the path his planet followed. To substitute a single rational God as the prime mover does not improve the situation. His plans, except as accomplished, remain inscrutable and there is the additional problem of saying why what is a rational path for Mercury is not a rational path for Venus.

The theories of Ptolemy, Copernicus, and Kepler, when stripped of alien addenda, do not provide mechanisms akin to those of Eudoxos. They provide alternative mathematical analyses of apparent celestial motions. Their explanations of these motions are formal or mathematical and not physical. Ptolemy and Copernicus both sought explanatory principles incorporating uniform, circular motions; Kepler broke with this tradition. There is some question whether Ptolemy's and Copernicus' analyses are to be taken as statements about what happens in nature, or instead as methods or devices for calculating the positions of celestial bodies in the sky. When Copernicus' book, *De Revolutionibus Orbium Coelestium*, appeared in 1543, the year of his death, it had a preface which claimed that the system presented was only a calculating device. It emerged later that Osiander, who supervised the printing of the book, wrote this preface. It is unlikely that Copernicus held the view it expressed. His disciple, Rheticus, for instance, denied that he did. Copernicus' painstaking refutation of arguments against the rotation of the Earth and for the rotation of the heavens *en bloc* constitutes internal evidence that he deemed himself to be writing about what is indeed the case in nature. The situation is less clear with respect to Ptolemy. In his theory of the Moon, the minimum distance of that body from the Earth was only half its maximum distance. Ptolemy would have recognized that this implied a variation in the apparent size of the Moon by a factor of two. He also knew that there is no marked, if any, variation in the apparent size of the Moon, apart from the well-known horizon-elevated Moon illusion which he recognized to be a psychological phenomenon and for which he provided a psychological explanation. As he attributed the variation in the brightness of the star-like planets to their variations in distance from the Earth as implied in the deferent-epicycle systems, it seems likely that he was simply dodging the uncomfortable implication by his theory of a marked variation in the apparent size of the Moon.

There can be no doubt about Galileo's rejection of the view that the Copernican system was just a calculating device or a manner of thinking about complex apparent celestial motions in simple theoretical terms. Cardinal Bellarmino put the view explicitly to him as a means of escape from the Church's displeasure. He declined to take that way out; indeed, ultimately he preferred to declare that he was wrong in supporting the Copernican system. If he ever said anything like the legendary, muttered: 'But it does move', it was perhaps: 'But it *really* does move'. Nor can there be any doubt that Kepler took his three laws to be assertions about the ways the planets actually move around the Sun and not merely convenient ways of analysing into simpler, regular motions their complex not patently regular apparent motions.

Setting aside this issue whether theories of the type, which we are discussing, are merely calculating devices or are conjectures about nature to be taken literally, there can be no disputing the differences between concrete, mechanical theories such as Eudoxos' (as it has been presented above) and abstract, mathematical theories such as Ptolemy's and Copernicus'. Newton's theory, unlike any of its predecessors, introduced a dynamic intrinsic to the events themselves. His notion of force may give us the conviction of a fuller explanation than that contained in Kepler's revision of Copernican theory because it is akin to the way our will or intention seems to act on our muscles when they move in fulfilment of our intention. It may be something one can make judgements about only by arguing back from the motions it is held to produce. However, Newton with his forces was in a stronger position than the Babylonians with their angels. They could conjecture what the angel's plan was, only by arguing back from the planet's observed path; further, their conjecture of the angel's plan enabled the inference of nothing but the planet's path. Ptolemy, of course, seems to have put himself largely in the same situation with his non-dynamic theory. He could conjecture what the radii of the deferent and epicycle were, and what their periods of revolution were, by arguing back from the planet's observed path; his conjecture, however, was not used to enable the inference of anything but the planet's path. Copernicus with his non-dynamic theory was not in like case. What he conjectured for one purpose, he could test otherwise; this enabled him to infer other things as well. Newton was in this position, too, with his dynamic theory. From his small number of dynamic principles a large network of implications could be traced. Interestingly Newton did not uncover them all. For at least two centuries his inheritors were still exploring the mine he bequeathed. His theory was thus a more comprehensive and more fertile theory as well as a more precise theory than those of his predecessors.

The testing of theories

We have seen that theories of the second type work not directly by the filling of a gap in factual knowledge but indirectly by implying matters of a factual sort. Likewise such theories are tested indirectly through the observation of what they imply and not directly by observation of what they state. Their central propositions taken by themselves do not ordinarily, if ever, explain and predict anything. Consequently such theories to be tested have to be taken in conjunction with other considerations which may be obtained from observation or by assumption. For example, the central propositions of Ptolemy's theory refer to epicycles borne on deferents 'focused' on the Earth. The additional considerations, also assumed, consist of ratios of the radii of epicycle and deferent, of the periods of revolution of epicycle and deferent, of the inclinations of the planes of these circles, of the amount of displacement of equants, and so on. The latter have to be written separately for each wanderer in the sky. Granted that the central propositions hold the promise of an accounting for the broad pattern of events to be explained, these other assumptions can be altered at will to give greater accuracy or to improve the accounting for minor detail. Epicycles of any desired radius and period of revolution can be added, their planes can be tilted to any desired degree and so on in order to yield the desired implications. Eudoxos and Ptolemy had no restraints upon them in such tinkering. Thus their auxiliary assumptions were completely *ad hoc* and their general theories were untestable except on the observed facts for the explanation of which they had been assumed. Their auxiliary assumptions taken in conjunction with the central parts of their theories had no other implications or at any rate no other implications which their authors took seriously. Thus any

73

attempt at independent testing was impossible. The whole procedure was circular. Whereas Ptolemy's theory and his assumed ratios of epicycles and deferents had no implications beyond the positions of the planets in the field of the fixed stars, Copernicus' introduction of the Earth's motion as a component of the apparent motions of the other planets transformed the situation. It required a given spacing of the planets in relation to the Sun, a matter open to estimation from independent direct observation. Further, as we have seen, it implied stellar parallaxes and, granted the finite velocity of light, the aberration of light. The increased range of explanatory power of a theory of this sort is important not merely because of the increased economy of assumption (each fact not requiring a separate assumption for its explanation) but also because values set in an *ad hoc* way for one explanatory purpose may be tested in other situations.

Eudoxos defended himself against independent testing by asserting that his spheres were invisible. In later times a great deal of importance was attached to establishing whether or not comets passed through the 'spheres' defined by the planetary orbits. If they did pass through the space supposedly occupied by the crystalline spheres, the credibility of the latter would have been undermined. However, a confirmed supporter of Eudoxos could have defended his hypothesis by asserting that the spheres were not only invisible but also penetrable by the kinds of material of which comets are made. Scientists have come to see that such in-built defences are too great a premium to pay for secured theorizing. Such defences do not aid the discovery of truth. They merely hinder the exposure of error.

Implication and consequence

It will be appropriate at this point to say a little more explicitly what is involved in implication or, more strictly speaking, formal implication. One assertion or set of assertions implies another when it can be shown that if one accepts the former one necessarily commits oneself to the latter. The former may be referred to as the *implicans* and the latter the *implicate*, that which implies and that which is implied respectively, though these terms have a wider meaning than the one specifically concerning us. Two or more simple propositions, *p*, *q*, *r*, asserted jointly constitute a compound proposition. One such manner of joint assertion is '*p* and *q*', a conjunctive proposition, e.g. it is raining and the bells are ringing. Another is: 'If p, then r', or more complexly: 'If p and q, then r', variously called conditional propositions, hypothetical propositions and implicative propositions. Consider the following cases:

(a) If I receive your letter by Tuesday, then I shall not telephone you.
(b) If I throw an object up into the air, then it will subsequently fall down to earth.
(c) If pieces of zinc are placed in hydrochloric acid, then hydrogen is given off.
(d) If kangaroos are marsupials and if marsupials deliver their young early in foetal development, then kangaroos deliver their young early in foetal development.

In each of them there is, or seems to be, some ground in the 'if p' or 'if p and q' or antecedent component for the assertion of the 'then r' or consequent component. If communication by letter has been achieved, then presumably there is no need for communication by telephone. Common experience has been summed up in: 'What goes up must come down.' Countless experiments in which zinc has been placed in hydrochloric acid have shown that hydrogen is given off. However, there are marked differences in the 'force' the antecedent has for the consequent in these three cases. The first must be a quite special case for the receipt of letters must almost as often prompt telephone calls as inhibit them. That is, it just so happens in this case that it will inhibit a telephone call. So that although the compound proposition (a) may be true, there is nothing of a general sort in its 'if p' which requires its 'then r'. There is more 'generality' in (b). Usually if objects are thrown up into the air, they fall down to earth. We can, of course, explain why this is so, though there is no hint of the explanation in (b). We would hardly ever throw a lighter-than-air object up into the air (indeed we would ordinarily just let it go if we want it to go up) and we hardly ever throw heavier-than-air objects with such vigour that they go into orbit or escape the earth's gravitational attraction. But even without this elaboration we recognize that when p in (b) is fulfilled, there is a high probability that r will also be fulfilled and a low probability that it will not. The situation is different in (c), as no instance of p's fulfilment has not been followed by r's fulfilment in human experience. So most of us would be willing to say that r following

75

p is a special consequence in (a), a frequent or common consequence in (b) and a universal (as far as we know) consequence in (c).

A little contemplation is likely to reveal that the consequence in (d) is also universal and that further there is no need to add 'as far as we know'. The implicate (r), in this fourth case follows from the conjunctive implicans $(p$ and $q)$ because of the form of the three propositions and not because of their content. If the two components in the implicans were true the implicate would necessarily be true. Because of the form of the three simple propositions, one would contradict himself were he to assert the first two and deny the third.

Consider these further examples:

(c') If pieces of gold are placed in hydrochloric acid, then hydrogen is given off.

(c") If pieces of zinc are placed in hydrochloric acid, then chlorine is given off.

(d') If frogs are amphibia and if amphibia have a heart with two ventricles, then frogs have a heart with two ventricles.

That gold is not attacked by hydrochloric acid, that the chlorine in hydrochloric acid combines with the zinc and is not given off, and that frogs have a heart with a single ventricle happen to be facts. Thus in (c') and (c") r does not follow from p' and r' does not follow from p. The universal consequence we found in (c) is a matter of fact; it is a factual consequence. It should be appreciated that the 'force' with which r' follows from p' and q' in (d') is no less than that with which r follows from p and q in (d). The content of the propositions and whether they are true or false has nothing to do with the 'force' of the argument. It is their form which provides this 'force'. The logical form of the argument is the same in (d) and (d'). Expanded, the argument is 'if all x's are y's and all y's are z's, then all x's are z's'. The implicate is a logical consequence of the joint implicans. This is what is meant by formal implication; it is something quite different from what has been called here factual consequence.

Factual consequences depend upon the factual or material truth of the conditional proposition, that is upon the implicate being a fact whenever the implicans is a fact. Formal implication depends in no way upon the factual or material truth of the component propositions. Hence, we say of a sound argument or a correct statement of an implication that it is valid, not that it is true. Validity is formal, truth is factual or material. Where the implicans formally implies the implicate, the former is said to be the premise (or if compound, premises) and the latter the conclusion. The conclusion follows from the premise(s) by virtue of the form of the propositions involved. When we speak of a theory implying certain facts (or their contraries) formal implication and not mere factual consequence is being referred to. False theories may formally imply conclusions which are true and/or conclusions which are false.

There are several formal or logical relations between propositions which are important for us to recognize. Two or more propositions may be compatible or incompatible. 'Mars is red' and 'Mars is not red' are clearly incompatible. One cannot consistently affirm them both together. 'Mars is red' and 'Mars is blue' are also incompatible. However, whereas one of the first pair must be true, neither of the second pair need be true. Mars is either red or not red, but Mars may be red or blue or yellow or white or some other colour. The first pair are contradictories—both cannot be true, but both cannot be false. The second pair are contraries—both cannot be true, but both may be false. The central propositions, asserting geocentricity and helio-centricity respectively, of Ptolemy's and Copernicus' theories are contraries and not contradictories. Both cannot be true but both may be false, for instance Brahe's theory could have been true. Hence, when Galileo with his observation of the phases of Venus proved Ptolemy's theory wrong, he did not prove Copernicus', or for that matter Brahe's, right.

Three relations of compatibility will be noted here. In the first, the relationship of independence, the affirmation or denial of one member of the pair commits one in no way to the affirmation or denial of the other. 'Mars is red' and 'Mars has two moons' are independent propositions. The acceptance or rejection of either formally involves one in no commitment to the other. In the second, the relationship of equivalence, the affirmation of either one involves one in the affirmation of the other, and the denial of either one involves one in the denial of the other. 'Kangaroos are never invertebrates' and 'invertebrates are never kangaroos' are equivalent propositions. In the case of independence, the truth or falsity of one proposition does not require the truth or falsity of the other; in the case of equivalence the truth of each requires the truth of the other and the falsity of each requires the falsity of the other. In between are cases where the truth of one requires the truth of the other but not vice versa, and the falsity of the other requires the falsity of the one but not vice versa. For example the affirmation of 'all kangaroos are plains-dwellers' requires the affirmation of 'some kangaroos are plains-dwellers' and the denial of 'some kangaroos are plains-dwellers' requires the denial of 'all kangaroos are plains-dwellers'. However, the denial of 'all kangaroos are plains-dwellers' does not require the denial of 'some kangaroos are plains-dwellers' (as the falsity of the former may consist in some kangaroos being plains-dwellers and some being hills-dwellers), and the affirmation of 'some kangaroos are plains-dwellers' does not require the affirmation of 'all kangaroos are plains-dwellers' (as some may be plains-dwellers and some not). This is the relation of implication which is not reciprocal. Where A implies B, and B implies A, that is, where there is reciprocal implication, A and B are equivalent.

These five formal relationships between propositions or sets of propositions are set out in Table 4.1.

Relation	If A is	then B must be	If B is	then A must be
1. A is contradictory to B	True	False	True	False
	False	True	False	True
2. A is contrary to B	True	False	True	False
	False	n.c.	False	n.c.
3. A is independent of B	True	n.c.	True	n.c.
	False	n.c.	False	n.c.
4. A is equivalent to B	True	True	True	True
	False	False	False	False
5. A implies B	True	True	True	n.c.
	False	n.c.	False	False

Table 4.1 Some logical relationships between pairs of propositions. *n.c.* symbolizes 'no commitment'.

An explanatory theory taken in conjunction with appropriate factual considerations implies some observable state of affairs. If observation shows that state of affairs to be contrary to what was implied then the theory or the other considerations or both must be false. If we have good grounds for accepting the other considerations the theory must be rejected. This is the way in which erroneous theories may be tested and eliminated. However, if observation corroborates what was implied, the theory is not proved to be true. This is what makes it possible for alternative theories to imply the same facts. Though we cannot in this way prove a theory, we may show it to be tenable. That is, it may be held consistently with matters of observation which, had they been other than they are, could have shown it to be false.

Quantitative facts

Our case studies revealed several types of facts as well as several types of theories. Observation may produce qualitative data, such as that venous blood is dark red and that arterial blood is bright red, or quantitative data, such as that Venus may be separated at the Earth by 48 degrees from the Sun and that Jupiter has twelve satellites. Quantitative data may consist of frequencies (answering the question 'how many?') or of amounts (answering the question 'how much?'). The crudest assessment of an amount is in terms of more, equal, and less. Among the most refined are assessments of length and weight expressed in terms of some established scale. Though the unit 'one foot' is arbitrary (it is said to be based on the length of the foot of one of the early English kings), measurement of length in terms of it is not. A foot-rule may be placed successively along an object allowing the length of that object to be assessed as so many times (or fractions of) the length of the rule. The additivity of lengths is a most important feature. Because we can add the length of the foot-rule to any other length, we know that the difference between 3 feet and 4 feet is equal to the difference between 11 feet and 12 feet, and we also know that 12 feet is twice as long as 6 feet. That is, we can justifiably compare intervals and speak of ratios of length.

In our next case study we will have occasion to speak of the softness or hardness of substances. The relative hardness of two substances may be established by trying to scratch one with the other. This establishes whether one substance is harder than, equal to or softer than the other. It does not, however, establish the magnitude of differences, except in crude terms such as this is easily (or not easily) scratched by that. Nevertheless, it is possible to select a series of substances, each of which is easily scratched by the next hardest, as standards in a scale of hardness. Mohs selected talc as the standard for the first grade of hardness, gypsum as the second, calcite as the third, fluorspar as the fourth, apatite as the fifth, orthoclase as the sixth, quartz as the seventh, topaz as the eighth, sapphire as the ninth,

and diamond as the tenth. Though this scale provides a useful stan-
dardized way of assessing and expressing the relative hardness of a
substance, it does not enable one to say, for instance, that the difference
between talc and gypsum is the same as the difference between topaz
and sapphire, or that gypsum is a fifth and apatite half as hard as
diamond. Hardness lacks additivity and so the best Mohs could do was
to produce an intensive scale of measurement. Scales of measurement
for additive properties such as length and weight, whether expressed in
feet or centimetres, whether in pounds or grammes, are extensive scales.

The measurement of temperature provides an interesting case.
One can detect relative differences in temperature by direct cutaneous
perception. It is convenient to have 'hot' and 'cold' taps so labelled,
but a finger held in the running water will also provide convincing
information about which is which. Many substances are seen to expand
with changes in temperature. Both mercury and coloured alcohol in
a very narrow tube with an attached bulb show marked changes of
length with changes of temperature. The Centigrade scale uses the
length of the column in the thermometer when pure water at atmospheric pressure freezes to set a zero point and the length when the
water boils to set a 100°C point. The Fahrenheit scale uses these two
events to set 32°F and 212°F points; this is an arbitrariness like using
the length of the king's foot or some fraction of a degree of latitude on
the Earth's surface as a unit of length. The value 0°C is not the coldest
anything can be; it is not 'no heat' as 0 feet or 0 centimeters is 'no
length'. Consequently 100°C is not twice as hot as 50°C, in the way
that 12 inches is twice as long as 6 inches. Hence the numbers we use
in the Centigrade and Fahrenheit scales in order to express temperatures may not be used to work out ratios of temperature. We may ask,
nevertheless, whether the difference between temperatures measured
as 40°C and 50°C are the same as the difference between temperatures
measured as 80°C and 90°C. We can say that the column of mercury
in the thermometer expands by the same linear amount between
40°C and 50°C as it does between 80°C and 90°C. We may therefore
answer our question in the affirmative if we have grounds for asserting
that the expansion of mercury is a linear function of change in temperature. Unless we move into a wider theoretical arena, the assertion
of this relationship is quite gratuitous. There are, however, theoretical
considerations, into which we cannot easily go here, which do provide
an adequate basis for the assertion. It is worth stressing that the basis

is theoretical rather than directly observational. To this extent 'facts' about temperatures already have a theoretical component in them derived from the way in which they are expressed. We shall find again and again that, though a distinction between theory and fact may be readily made in general terms, specific instances of our knowledge are mixtures, in varying proportions, of theory and fact.

Another interesting case of a measuring scale, is that used to express the apparent brightness of the stars. Hipparchos selected some twenty of the brightest stars and labelled them as being of the first magnitude. He graded the stars of less brightness into five other classes, the just visible being of the sixth magnitude. Ptolemy introduced steps within the magnitudes. This scale may be illustrated by reference to that easily identified constellation *Orion* (*see* Fig. 4.1). Betelgeuse, *alpha Orionis*, the red star at Orion's right shoulder and Rigel, *beta Orionis*, the bluish star at Orion's left foot, are first magnitude stars. *Gamma*, at the left shoulder, *kappa*, at the right foot, and *zeta*, *epsilon* and *delta*, the three brightest stars in Orion's belt are second magnitude stars. *Eta Orionis*, at right angles to the top of the belt and at Orion's left hip, *lambda*, the brightest star at Orion's head, *iota*, the brightest star in Orion's sword, and *tau*, near Rigel, perhaps at Orion's left ankle, and *pi* 3, in the middle of the lion's skin held in Orion's left hand, are stars of the third magnitude. *Phi* 1 and 2, the second and third brightest stars at Orion's head, four stars, *omega*, *psi*, 25 and 32, scattered about Orion's chest, together with *mu*, which marks the beginning of Orion's upraised right arm, are stars of the fourth magnitude. The faintest visible stars at Orion's head, in his belt and in his sword, are stars of the sixth magnitude.

The brightness of a star, one may reasonably presume, is a direct function of the amount of visible radiant energy it gives out (subject to an absence of obscuring matter between it and us) and an inverse function of its distance. These days the astronomers have devices such as photoelectric cells which measure the amount of radiant energy of given wave-lengths reaching us from any given star. Is the difference between the amount of light reaching us from stars of the first and the second magnitude the same as the difference between the amount of light reaching us from stars of the fifth and the sixth magnitude? Is the amount of light reaching us from a first magnitude star six times greater than that reaching us from a sixth magnitude star and twice as great as that from a third magnitude star? 'No' is the answer to all these questions.

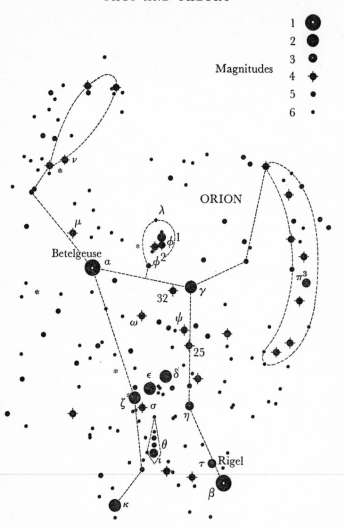

Fig. 4.1 The constellation Orion showing stars of the several visible magnitudes.

The magnitudes introduced by Hipparchos are graded in terms of apparent brightness and this happens not to be in a rectilinear relationship to the amount of light reaching the eye from the object being judged for brightness. Weber showed in the early nineteenth century that visible radiant energy needs to increase not by a given amount but by a given proportion, namely about a sixtieth, to be visibly different. The general principle may be appreciated in the recognition that whereas a lighted match may add perceptibly to the illumination given by a candle, it will make no apparent difference whatsoever to the perceptible illumination afforded by a searchlight. The light of a match is a modest fraction of the light of a candle but is only a negligible fraction of the light radiated by a searchlight. As Fechner later showed differences in apparent brightness vary approximately as the logarithm of the radiant energy. Stars differing by one magnitude stand approximately in the ratio 2.5 : 1 in terms of radiant energy received by the eye of the observer. The radiant energy received from a first magnitude star is a hundred times greater than that received from a sixth magnitude star, namely $(2.512\;..)^5$. Once this relationship between apparent brightness and amount of radiant energy received by the observer was recognized, it became possible to extrapolate the apparent scale to those very bright stars which are brighter than most other stars graded by Hipparchos as of the first magnitude and to stars invisible to the naked eye. Thus Sirius, the apparently brightest fixed star, has a magnitude of −1.58. The nearest star to the Earth, after the Sun, is *proxima Centauri* which has a magnitude of 10.5; it is thus well beyond the limits of detection by the naked eye. Table 4.2 sets out the apparent magnitudes of the twenty-two brightest stars.

Star	Apparent magnitude	Star	Apparent magnitude
Sirius	−1.58	Betelgeuse	0.92
Canopus	−0.86	*alpha Crucis*	1.05
alpha Centauri	0.06	Aldebaran	1.06
Vega	0.14	Pollux	1.21
Capella	0.21	Spica	1.21
Arcturus	0.24	Antares	1.22
Rigel	0.34	Fomalhaut	1.29
Procyon	0.48	Deneb	1.33
Achernar	0.60	Regulus	1.34
beta Centauri	0.86	*beta Crucis*	1.50
Altair	0.89	Castor	1.58

Table 4.2   Apparent magnitudes of the brightest twenty-two stars.

Though apparent brightnesses are not ordinarily additive in reality (amounts of radiant energy are, but that is another matter), they are divisible in judgement. A light may be adjusted until it appears half as bright or twice as bright as a standard light. Thus a scale possessing many of the features of an extensive scale may be set up, even though the property of seen or judged brightness is strictly speaking intensive.

Measurement theorists are prone to adopt a too purist view in which only extensive properties are admitted as literally measurable. The prescriptive philosopher rather than the expedient scientist is the promoter of such a view. We must recognize that though science cannot afford to adopt as a guiding principle the dictum that anything goes, it equally cannot afford to rule out anything which does work even if not perfectly.

Facts not independent of theories

It has been convenient above to speak of alternative theories as explaining the same set of facts. We have seen how the heliofocal theory explained facts other than the successive directions of the wanderers, for example the relative distances of the members of the Solar System from the Sun, the phases of Mercury and Venus (and for that matter less marked phases of Mars) and stellar parallaxes. Consequently the facts it explains are not the same set as those explained by the Eudoxan and the Ptolemaic theories. At the very least it is a more extensive set. One may go further and say that not only has the original set been extended or added to but also it has in various ways been altered by the new way the theory requires them to be interpreted. For instance, in Copernican theory the fixed stars are no longer regarded, as they were in Eudoxan theory, as being equidistant. From this it follows that their apparent brightnesses are not determined solely by their absolute brightnesses. The ways in which theories affect what we take to be the facts will become even clearer in the two following case studies.

Part Three

THE PERIODIC TABLE OF ELEMENTS

5 CASE STUDY

Chemistry began to take its modern shape in the century and a half between Robert Boyle's *Sceptical Chymist* (1661) and John Dalton's *A New System of Chemical Philosophy* (1808-10). The basic gas laws were formulated. New elements such as hydrogen, nitrogen, and oxygen were discovered. Dalton in his atomic theory brought several ideas, which had been emerging, to a state of clarity. He made a clear distinction between elements and compounds, the latter being combinations of the former. He showed that the elements combined in definite proportions and not in any proportions whatsoever. He assumed that in combining, the atoms of elements were not changed but were rearranged. These notions may be illustrated by reference to hydrogen, oxygen and water as we now understand their relations. Hydrogen and oxygen can be made to combine to form water. Water can be broken into hydrogen and oxygen; water is thus a compound. As hydrogen and oxygen cannot be broken by chemical means into anything else, they are elements. Their combining and separating suggest that water is a re-arrangement of the units (atoms) of hydrogen and oxygen, and not a transformation of them. It was eventually established that one part of hydrogen to eight parts of oxygen, by weight, are always involved in the formation of water. Once it became possible on certain theoretical bases to estimate the number of atoms involved in compounds, it became possible to establish what were called atomic weights but are now more correctly called atomic masses. Thus if two atoms of hydrogen (H) combine with one atom of oxygen (O) to form water (H_2O) then the atomic weights of hydrogen and of oxygen must stand in the ratio of one to sixteen.

By establishing the proportions in which elements typically or most readily combine, a value termed valence may be obtained. Hydrogen and oxygen typically combine in the proportions of two atoms of hydrogen to one atom of oxygen, so do lithium (Li) and oxygen, and sodium (Na) and oxygen. Hydrogen, lithium and sodium are consequently said to have a valence of 1 and oxygen a valence of 2. Carbon (C) combines with oxygen to form carbon monoxide (CO) and carbon dioxide (CO_2). Taking the higher of these two oxides, that is CO_2, carbon is said to have a valence of 4 as one atom of carbon combines with two atoms of oxygen, each of which has a valence of 2. Nitrogen (N) combines with hydrogen in the proportion of one to three to give ammonia (NH_3); thus nitrogen has a valence of 3.

The extending identification of elements and the fuller knowledge of their properties led in the nineteenth century to the recognition of several families of elements. A striking case is presented by the halogens, a group of elements readily forming white crystalline salts. The family name derives from the Greek words *hals*, salt, and *genos*, origin. Omitting a more recently recognized member, the halogens are fluorine (F), chlorine (Cl), bromine (Br), and iodine (I); the first two are gases, the third a liquid and the fourth a volatile solid. The salts they form all have comparable formulae, for example, the sodium (Na) salts are NaF, NaCl (common salt), NaBr, and NaI, and the aluminium (Al) salts are AlF_3, $AlCl_3$, $AlBr_3$, and AlI_3. All form simple compounds with hydrogen, HF, HCl, HBr, and HI which dissolve in water to form acids. It will be noted that all these compounds indicate that these four halogens have the same valence. In addition to properties such as those cited in which they are alike, the halogens have other properties which vary in an orderly manner amongst them. Three of these which are set out in Table 5.1 are parallelled by the readiness with which the halogens combine with hydrogen, fluorine doing so most readily and iodine least.

Halogen	Atomic weight	Melting point	Boiling point
Fluorine	19.00	−223.0°C	−187.0°C
Chlorine	35.46	−101.6°C	−34.6°C
Bromine	79.17	−7.2°C	58.8°C
Iodine	126.91	113.5°C	184.0°C

Table 5.1   Some properties of the halogens manifesting an orderly variation.

Another early recognized family is the set of alkali metals. The commonest are sodium and potassium (K). The compounds of these

1. hydrogen	8. fluorine	15. chlorine	22. cobalt nickel	29. bromine	36. palladium	42. iodine	50. platinum iridium
2. lithium	9. sodium	16. potassium	23. copper	30. rubidium	37. silver	44. caesium	53. thallium
3. beryllium	10. magnesium	17. calcium	25. zinc	31. strontium	38. cadmium	45. barium vanadium	54. lead
4. boron	11. aluminium	19. chromium	24. yttrium	33. cerium lanthanum	40. uranium	46. tantalum	56. thorium
5. carbon	12. silicon	18. titanium	26. indium	32. zirconium	39. tin	47. tungsten	52. mercury
6. nitrogen	13. phosphorus	20. manganese	27. rsenic	34. didymium molybdenum	41. antimony	48. niobium	55. bismuth
7. oxygen	14. sulphur	21. iron	28. selenium	35. rhodium ruthenium	43. tellurium	49. gold	51. osmium

Table 5.2 Newlands' table of elements. The number preceding each substance gives the order of its atomic weight as known to Newlands. Didymium is now known not to be an element but a naturally occurring mixture of two elements, praseodymium and neodymium, with atomic weights near to each other (140.92 and 144.27 respectively). It should be noted that Newlands did not arrange the elements in strict order of atomic weight as known to him.

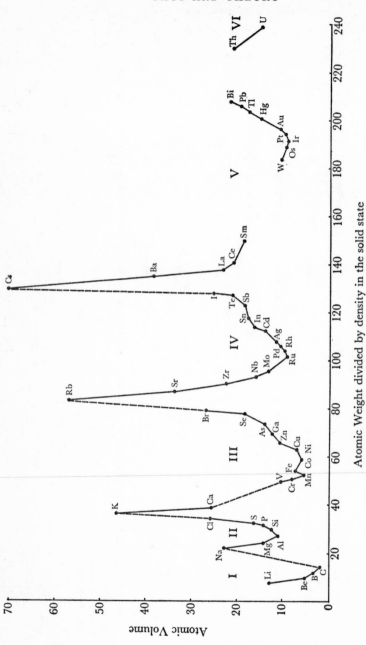

Fig. 5.1 Lother Meyer's graph showing the periodic relation of the atomic volume of the element to the atomic weight divided by the density of the element in the solid state.

two elements with carbon and oxygen, sodium carbonate or washing soda (Na_2CO_3) and potassium carbonate (K_2CO_3) are found in abundance in plant ashes for which the Arabic word is *al-qili*. Lithium (from the Greek *lithos*, a stone) is rare though traces are present in most rocks. Rubidium (Rb) and caesium (Cs) are very rare. All these alkali metals are lustrous in appearance and are good conductors of heat and electricity—these are their metallic properties. They are relatively soft metals and are highly malleable and ductile. They form hydroxides which are markedly alkaline or basic, a property opposite to being acidic. Caustic soda (NaOH) is the most widely known of these hydroxides. The others, LiOH, KOH, RbOH, and CsOH have analogous formulae and similar properties. In some other respects the alkali metals show an orderly variation; their atomic weights, melting points and boiling points are parallel as in the case of the halogens.

The atomic weight of an element in a family was usually found to be close to the average of the atomic weights of its next lighter and next heavier family members. Thus sodium has an atomic weight of 22.99 which is close to the average (23.02) of the atomic weights of lithium (6.94) and of potassium (39.10). This suggested some regularity in the spacing of the atomic weights of the elements. Such a regularity in other respects was also suggested when the elements were arranged in order of atomic weights.

The British chemist Newlands, in 1865, arranged the elements as then known in approximate order of atomic weight (*see* Table 5.2). He set the elements out in eight columns of seven members or, in six cases where the atomic weights seemed to be the same, pairs of members. In the first two columns, Newlands had the elements in order of their known atomic weights; in the next five columns he had three one-place transpositions in order to bring out better the pattern he wished to reveal, but in the eighth column he almost ignored the order of atomic weights. He was, of course, handicapped by uncertainty about the atomic weights of some elements and in other cases the information he had was grossly in error: thus uranium was thought to be much lighter than antimony, and indium lighter than arsenic. He drew attention to the way in which like elements occurred in the same rows. Thus the halogens, with some other elements, occupy the first row; the alkali metals, with some others, the second row; the alkali earths, with some others, the third row. Lithium, sodium and potassium form the beginnings and ends of two successive octaves, rubidium and

Group →	I	II	III	IV	V	VI	VII	VIII
Higher oxides and hydrides	R_2O	RO	R_2O_3	RO_2 H_4R	R_2O_5 H_3R	RO_3 H_2R	R_2O_7 HR	RO_4
1	H(1)							
2	Li(7)	Be(9.4)	B(11)	C(12)	N(14)	O(16)	F(19)	
3	Na(23)	Mg(24)	Al(27.3)	Si(28)	P(31)	S(32)	Cl(35.5)	
4	K(39)	Ca(40)	—(44)	Ti(48)	V(51)	Cr(52)	Mn(55)	Fe(56), Co(59), Ni(59), Cu(63)
5	[Cu(63)]	Zn(65)	—(68)	—(72)	As(75)	Se(78)	Br(80)	
6	Rb(85)	Sr(87)	?Yt(88)	Zr(90)	Nb(94)	Mo(96)	—(100)	Ru(104), Rh(104), Pd(106), Ag(108)
7	[Ag(108)]	Cd(112)	In(113)	Sn(118)	Sb(122)	Te(125)	I(127)	
8	Cs(133)	Ba(137)	?Di(138)	?Ce(140)	—	—	—	
9	—	—	—	—	—	—	—	
10	—	—	?Er(178)	?La(180)	Ta(182)	W(184)	—	Os(195), Ir(197), Pt(198), Au(199)
11	[Au(199)]	Hg(200)	Tl(204)	Pb(207)	Bi(208)	—	—	
12	—	—	—	Th(231)	—	U(240)		

Table 5.3 The 1872 version of Mendeleyev's periodic table of elements. (Reprinted by permission from G. Holton

caesium are two double octaves further on. Likewise fluorine and chlorine form an octave, with bromine and iodine double octaves further on. Newlands felt the parallel with the tonal scale of octaves to have some special weight. The later discovery of the inert gases removed this rather Pythagorean feature, replacing the sets of eight with sets of nine.

Without seeking so strictly an ordered pattern, the German chemist Lothar Meyer, in 1869, showed that there was a periodicity in the atomic volume of elements when they were arranged in order of atomic weight (*see* Fig. 5.1). In the same year the Russian Dmitri Ivan Mendeleyev produced the first version of his famous periodic table of elements. He arranged the elements by rows in order of atomic weight and by columns in accordance with valence. This dual classification involved leaving blanks in some cells of the table as may be seen in the 1872 version which is presented in Table 5.3 In this table, Mendeleyev has eight groups of elements, the members of each group having the same valence, and twelve rows or series running across the groups. Hydrogen is the sole entry in series 1, being placed in group I as its higher oxide is H_2O (water) which indicates its valence to be 1. Lithium is the next entry being placed in series 2 and group I, its higher oxide being Li_2O, which indicates that its valence also is 1. Following it in series 2 are beryllium in group II, its oxide being BeO; boron in group III, its oxide being B_2O_3; carbon in group IV, its higher oxide being CO_2; nitrogen in group V, its highest oxide being N_2O_5; oxygen in group VI, its higher hydride being H_2O and fluorine in group VII, its hydride being HF. Series 2 is blank in the eighth column. The next element by atomic weight known at that time was sodium, which Mendeleyev placed at the beginning of series 3 under lithium the immediately lighter alkali metal. After it came magnesium in group II like beryllium in the series above, both being alkaline earths. And so on. The alkali metals were placed with hydrogen in group I, the alkaline earths together with zinc (Zn), cadmium (Cd), and mercury (Hg) were in group II, and the halogens together with manganese (Mn) were in group VI. The other groups contained families with at least noticeable if not striking family resemblances. The clusters of entries in group VIII are metals—iron (Fe), cobalt (Co), nickel (Ni), and copper (Cu) in series 4; ruthenium (Ru), rhodium (Rh), palladium (Pd), and silver (Ag) in series 6; and osmium (Os), iridium (Ir), platinum (Pt), and gold (Au) in series 10. The

93

fourth member of each of these clusters was also more tentatively assigned by Mendeleyev to group I, indicating that he was not quite sure of the details of the broad regularity he had found.

A further evidence of his uncertainty was contained in his suggestion that greatest resemblances in a group were to be found in alternative series. If one considers group II, zinc in series 5, cadmium in series 7 and mercury in series 11 resemble one another more than they resemble the alkaline earths occupying series 2, 3, 4, 6 and 8. (*Note* that magnesium in series 3 does not fit the alternate series suggestion.) Again in group VII, fluorine in series 2, chlorine in series 3, bromine in series 5, and iodine in series 7 are more alike than manganese in series 4 is like any of them. (*Note*, however, that fluorine does not fit the alternate series suggestion.)

Of special interest to us are the cells marked with a dash and a parenthetical atomic weight. Consider series 4. Calcium is properly located in group II its oxide being CaO. The next heavier element, titanium (Ti), belongs not in group III but in group IV as its oxide is TiO_2. Mendeleyev, therefore, supposed there to be an element of atomic weight 44 (midway between calcium, 40 and titanium, 48) which formed an oxide X_2O_3, which was basic but not very active, which was a nonvolatile metal and so on. Such an element would fill the gap in series 4, group III. Its properties were inferred by interpolation from those of its known neighbours. Likewise for a gap in series 5 in group III he assumed an unobserved element of atomic weight 68 forming an oxide Y_2O_3 and being a metal more volatile than aluminium. In the same series in group IV he assumed another element of atomic weight 72, forming an oxide ZO_2, and so on. The three elements were observed and named respectively scandium by Nilson and Cleve in 1880, gallium by Lecoq de Boisbaudran in 1875, and germanium by Winckler in 1885. The similarity of Mendeleyev's predictions and Winckler's observations is impressive as Table 5.4 shows.

These were signal confirmations of Mendeleyev's assumption that many properties of elements could be inferred, by interpolation, from the properties of their neighbours. He also maintained that where a property of an observed element could not be established precisely by observation, similar inferences by interpolation could bring greater certainty. He urged this particularly in respect of atomic weights which were uncertain for a number of elements. It is worth remarking upon the paradox here. Mendeleyev arranged the elements in the table

94

	Mendeleyev's prediction	Winckler's finding
Atomic weight	72	72.6
Specific gravity	5.5	5.36
Oxidization formula	ZO_2	GeO_2
Specific gravity of oxide	4.7	4.7
Solubility of salts in water	high	high
Chloride—formula	ZCl_4	$GeCl_4$
—state	liquid	liquid
—boiling point	About 90° C	83° C
—specific gravity	1.9	1.887

Table 5.4   Comparison of some of Mendeleyev's predictions with analogous findings by Winckler on germanium.

in order of their atomic weight, and then he suggested that the arrangement could be used to establish atomic weight with greater precision in certain cases. The paradox is reduced a little if we think of a row of posts in a fence lying approximately in a straight line. Provided we can establish the general direction of the fence, uncertainties about the positions of particular posts may reasonably be reduced by reference to the main line of the fence. Nevertheless it proved to be the case here that some posts in this fence were well out of line. The most striking anomaly is provided by tellurium and iodine, believed by Mendeleyev to have atomic weights of 125 and 127 respectively. They fitted nicely in groups VI and VII respectively. Yet we now know that the atomic weight of tellurium is 127.61 and that of iodine is 126.91. Mendeleyev put them in the right groups though the order of their atomic weights require them to be transposed. We shall examine this matter further below.

With the advance of empirical knowledge, it became necessary by the second decade of the twentieth century to reshape the periodic table (*see* Table 5.5). An additional group was added for the inert gases, helium, neon, argon, krypton, etc. These are termed inert because they were believed not to combine with other elements; it is now known that some of them do form a few compounds. In the arrangement displayed in Table 5.5, helium is added to hydrogen in period, or series, 1 being placed in an additional 'zero' group. Periods

95

Group	I	II	III	IV	V	VI	VII	VIII	0
Period 1	H 1.008								He 4.0
Period 2	Li 6.9	Be 9.0	B 10.8	C 12.0	N 14.0	O 16.0	F 19.0		Ne 20.2
Period 3	Na 23.0	Mg 24.3	Al 27.0	Si 28.1	P 31.0	S 32.1	Cl 35.5		Ar 39.9
Period 4 (long)	K 39.1 / Cu 63.6	Ca 40.1 / Zn 65.4	Sc 45.1 / Ga 69.7	Ti 48.1 / Ge 72.6	V 51.0 / As 75.0	Cr 52.0 / Se 79.2	Mn 54.9 / Br 79.9	Fe 55.8 Co 58.9 Ni 58.7	Kr 82.9
Period 5 (long)	Rb 85.4 / Ag 107.9	Sr 87.6 / Cd 112.4	Yt 88.9 / In 114.8	Zr 91.0 / Sn 118.7	Nb 93.1 / Sb 121.8	Mo 96.0 / Te 127.5	I 126.9	Ru 101.7 Rh 102.9 Pd 106.7	Xe 130.2
Period 6 (long)	Cs 132.8 / Au 197.2	Ba 137.4 / Hg 200.6	La 138.9 / Tl 204.4	Hf 178.6 / Pb 207.2	Ta 181.5 / Bi 209.0	W 184.0 / Po ?	Re ? / —	Os 190.8 Ir 193.1 Pt 195.2	Rn 222.0
Period 7	—	Ra 226.0	Ac ?	Th 232.1	—	U 238.2	—		

Table 5.5 The periodic table as it was often set out in the early twentieth century. It was recognized that lanthanum (La) was followed in its cell by fourteen other rare earths of very similar properties: those known by Mendeleyev gave him difficulties (*see* his entries Di, Ce, Er and La in Table 5.3). It has since been generally agreed that thorium (Th), uranium (U) and a number of other recently-established elements should be placed after actinium (Ac) in its cell.

2 and 3 are the same as Mendeleyev's series 2 and 3 with neon added in the zero group in the former and argon in the latter. Periods 4, 5, and 6 are formed by combining Mendeleyev's series 4 and 5, 6 and 7, and 8 and 10 respectively to form 'long' periods. Each of these 'long' periods has an inert gas added at the end of it. Within these 'long' periods the symbols are placed in the cells so as to emphasize that the first two elements and the last five (excluding the terminal inert gas) are members of the families defined by the first seven members of the 'short' periods 2 and 3. The members of group VIII in the 'long' periods had come to be called transition elements as they fell between the 'octaves' formed by the members before and after them in the period. Later the elements not members of the families defined by the members of periods 2 and 3, namely scandium to zinc in period 4, yttrium to cadmium in period 5 and lanthanum (with its companion rare earths) to mercury in period 5, were also called transition elements.

It will be noted that several elements are placed in the columns of this later table not in accordance with the order of their atomic weights. Thus argon, 39.9, is placed before potassium, 39.1; cobalt, 58.9 before nickel, 58.7; and tellurium, 127.5, before iodine, 126.9. In these cases family resemblances were deemed to outweigh the strict order of atomic weight. It became customary to speak of the ordinal number of the element in the table as its atomic number. Thus the atomic number of hydrogen is 1, helium 2, lithium 3, and so on. In the early years of this century Rutherford suggested that atoms were composed of a massive positively charged nucleus surrounded by a less massive (and more 'open') negatively charged outer region. His indirect measurements of the nuclear charges of several elements indicated that they corresponded to the atomic numbers. That this is a precise relationship has since been confirmed.

From this point onwards it will be more appropriate to use the term atomic mass. For hydrogen the atomic mass approximates the atomic number, for helium it is double, for lithium and the elements beyond it is more than double, the excess tending to become greater as we proceed through the elements. Radium has an atomic mass 2.57 times its atomic number (88, allowing for the fourteen rare earths following lanthanum in group III). There is a hint of some pattern in this also but it is less clear than relationship between atomic number and nuclear charge.

At this stage of its history the table had several puzzling features;

97

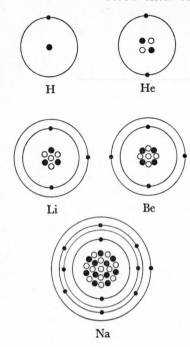

H   He

Li   Be

Na

Fig. 5.2 Diagrams showing the particles in the nuclei and in the shells of the atoms of five elements according to Bohr's theory. The larger filled circles represent protons, the open circles neutrons and the smaller filled circles electrons.

further, though revealing an unmistakeable pattern, it was without a rationale. It had enough regularity to suggest there was some underlying principle and to enable predictions that subsequent observations corroborated, yet it was still essentially *ad hoc*. It was still a 'brute' rather than a 'rational' regularity.

On grounds largely unrelated to these puzzles and unanswered questions about the periodic table, the Danish physicist Nils Bohr (1885-1962), following the earlier suggestions of Rutherford, developed a theory about the internal structure of atoms which provided a rationale for the table in respect both of its general form and of many, if not all, of its details. Though Bohr's theory has undergone subsequent modification, a simplified statement of its early form should suffice here. Bohr thought of atoms as being made up of a number of fundamental particles. Three such particles will be mentioned here, the proton, the electron and the neutron. The proton was assumed to carry a positive electrical charge, the electron a negative charge and the neutron to be electrically neutral. The proton and the neutron were assumed to be about equally massive and the electron to be distinctly less massive. Their masses are assessed as being, using approximate values, as follows:

proton	1.0076 *amu*
neutron	1.0090 *amu*
electron	0.0005 *amu*

The atomic mass unit, *amu*, is $\frac{1}{12}$th the mass of a carbon atom (strictly the most common carbon isotope, carbon 12).

Bohr thought of the atom as consisting of a nucleus, with one or more electrons in orbit around the nucleus. Bohr supposed the hydrogen atom to have one proton in its nucleus and one electron in orbit around it. Thus the atomic mass of hydrogen, 1.008, is accounted for. Helium is supposed to have a nucleus of two protons and two neutrons, and to have two electrons in orbit around the nucleus. Thus, allowing for loss of mass through the expenditure of energy in packing the nucleus, the atomic mass of helium, 4.003, is accounted for. In like manner, lithium has three protons, four neutrons and three electrons; beryllium, four protons, five neutrons and four electrons; sodium, eleven protons, twelve neutrons and eleven electrons. These assumed states of affairs are represented in Fig. 5.2. The number of protons corresponds to the atomic number of the element—hydrogen 1, helium 2, lithium 3, beryllium 4, sodium 11. The sum of the protons and neutrons approximates the atomic mass of the element—hydrogen, 1 particle in the nucleus, atomic mass 1.008; helium, 4 particles in the nucleus, atomic mass 4.003; lithium, 7 particles in the nucleus, atomic mass 6.940; beryllium, 9 particles in the nucleus, atomic mass 9.013; sodium, 23 particles in the nucleus, atomic mass 22.00, in each citing present values of atomic mass.

Bohr assumed that the electrons moved around the nucleus in one or other of a series of orbits or 'shells' of different distance from the nucleus and of different energy. He deemed these energies to be quantal, that is, having one or other of a series of discrete values without intermediate values. As electrons and protons are opposite in charge they should attract one another and draw together unless, for instance, the former are moving with high velocity around the latter. (The planets would move into the Sun under gravitational attraction were they not in motion around it.) But in other cases electric charges in motion give off energy. Were electrons to do this, they would collapse into the nucleus just as the planets would collapse into the Sun were

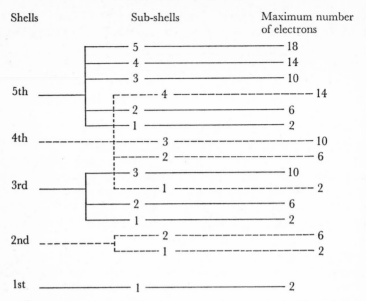

Fig. 5.3 Diagram showing the energy levels (sub-shells) in the several electron shells. The maximum number of electrons at each level is recorded.

they to lose their inertial motion. In certain circumstances electrons do give off energy but only in discrete amounts, i.e. when they drop from one energy level, or one sub-shell within a shell, to another below it.

Beyond the first shell in which the one electron of hydrogen and the two electrons of helium were assumed to orbit, the shells were assumed to have two or more sub-shells or energy levels. The third sub-shell of the third shell was deemed to be more distant from the nucleus than the first sub-shell of the fourth shell. A similar overlap was assumed to occur with the fourth and the fifth shells and with the fifth and the sixth shells. Bohr further assumed that there was a maximum number of electrons for each shell and sub-shell. Thus the first shell was said to be full when it had two electrons, that is, in helium. The first sub-shell of the second shell was full when it had two electrons and the second sub-shell when it had six electrons. Thus the elements in order from lithium to neon all have a 'helium core', that is, two electrons in the first shell. Lithium adds an electron in the first sub-shell of the second shell and beryllium adds two; boron has in addition one electron in the second sub-shell; carbon two; nitrogen three; oxygen four; fluorine five; and neon six. In both helium and neon the outermost shell is full and on this account the element is inert.

Hydrogen has one vacancy in its only shell and oxygen has two in its outermost shell. The two elements may combine by the electrons in the outer shells of two atoms of hydrogen filling the two vacancies in the outer shell of oxygen. Thus hydrogen combines with oxygen in the proportion of two atoms to one to form water (H_2O). Likewise the electrons of three hydrogen atoms are required to fill the three vacancies in the outer shell of a nitrogen atom in order to form ammonia (NH_3). Fluorine has one vacancy so it combines in equal proportions with hydrogen to form hydrogen fluoride (HF). It will be seen that this explains the general fact of valence as well as the valence of individual elements.

The members of the third period have a neon 'core', that is, two electrons in the first shell, two in the first sub-shell, and six in the

Shell	1	2		3			4			
Level	1	1	2	1	2	3	1	2	3	4
Ar	2	2	6	2	6					
K	2	2	6	2	6	—	1			
Ca	2	2	6	2	6	—	2			
Sc	2	2	6	2	6	1	2			
Ti	2	2	6	2	6	2	2			
V	2	2	6	2	6	3	2			
.	.	.	.	.	.	.	.			
.	.	.	.	.	.	.	.			
Zn	2	2	6	2	6	10	2			
Ga	2	2	6	2	6	10	2	1		
Ge	2	2	6	2	6	10	2	2		
As	2	2	6	2	6	10	2	3		
Se	2	2	6	2	6	10	2	4		
Br	2	2	6	2	6	10	2	5		
Kr	2	2	6	2	6	10	2	6		

Table 5.6 Electron distribution of some members of period 4; only four, scandium, titanium, vanadium and zinc, of the transition elements are shown.

FACT AND THEORY

second sub-shell of the second shell. Sodium (*see* Fig. 5.2), has, in addition, one electron in the first sub-shell of the third shell and magnesium two in that sub-shell. Aluminium adds an electron in the second sub-shell of the third shell, silicon two, phosphorus three, sulphur four, chlorine five and argon six. As argon has its outermost sub-shell full, it is inert.

The members of the fourth, a 'long', period have an argon 'core'. Potassium adds one electron not in the third sub-shell of the third shell but in the first sub-shell of the fourth shell; calcium has two in that sub-shell. The next run of elements in the period, those which have cousinly, rather than brotherly, resemblances to the main members of their groups, progressively fill the third sub-shell of the third shell. Thus scandium has one electron in that sub-shell, titanium two, vanadium three, and so on to zinc with ten. The corresponding members of the other 'long' periods also belatedly fill the last sub-shell of the next to the outermost shell. They are now added to the members of group VIII as transition elements. The remaining members of the fourth period progressively fill the second sub-shell of the fourth shell, as shown in Table 5.6. Krypton (Kr), which has all its sub-shells filled, is inert.

The theory also explains the fact that the members in each period show a gradation of properties, for example from metallic to non-metallic, from those forming basic to those forming acidic compounds, though the demonstration of that is rather too technical for this exposition.

Shells	1	2		3			4				5			
Levels	1	1	2	1	2	3	1	2	3	4	1	2	3	4
B	2	2	1											
Al	2	2	6	2	1									
Sc	2	2	6	2	6	1	2							
Ga	2	2	6	2	6	10	2	1						
Y	2	2	6	2	6	10	2	6	1	—	2			
In	2	2	6	2	6	10	2	6	10	—	2	1		

Table 5.7 Electronic configurations of elements originally placed in group III.

102

Period	1a	2a	3b	4b	5b	6b	7b	8	8	8	1b	2b	3a	4a	5a	6a	7a	0
I	1 H																	2 He
II	3 Li	4 Be											5 B	6 C	7 N	8 O	9 F	10 Ne
III	11 Na	12 Mg											13 Al	14 Si	15 P	16 S	17 Cl	18 Ar
IV	19 K	20 Ca	21 Sc	22 Ti	23 V	24 Cr	25 Mn	26 Fe	27 Co	28 Ni	29 Cu	30 Zn	31 Ga	32 Ge	33 As	34 Se	35 Br	36 Kr
V	37 Rb	38 Sr	39 Y	40 Zr	41 Nb	42 Mo	43 Tc	44 Ru	45 Rh	46 Pd	47 Ag	48 Cd	49 In	50 Sn	51 Sb	52 Te	53 I	54 Xe
VI	55 Cs	56 Ba	*	72 Hf	73 Ta	74 W	75 Re	76 Os	77 Ir	78 Pt	79 Au	80 Hg	81 Tl	82 Pb	83 Bi	84 Po	85 At	86 Rn
VII	87 Fr	88 Ra	**															

Lanthanum series:

*	57 La	58 Ce	59 Pr	60 Nd	61 Pm	62 Sm	63 Eu	64 Gd	65 Tb	66 Dy	67 Ho	68 Er	69 Tm	70 Yb	71 Lu

Actinium series:

**	89 Ac	90 Th	91 Pa	92 U	93 Np	94 Pu	95 Am	96 Cm	97 Bk	98 Cf	99 Es	100 Fm	101 Md	102 No

Table 5.8   The periodic table as it is now customarily set out. The numeral above the symbol for each element gives its atomic number. The lanthanum series, 57-71, is placed in the cell between barium and hafnium, and the actinium series, 89-102, in the cell after radium.

Thus the theory explains both family resemblances and individual characteristics. It explains not only why hydrogen, lithium, sodium and so on have the same valence, but also why they are monovalent. It has not yet been possible to deduce from it every property of all the elements but it is not improbable that such deductions will be made.

The theory has also led to a change in the way the periodic table is set out. The electronic configurations of the transition elements in the middle of each of the three 'long' periods involve the progressive filling of the outer (overlapping) sub-shell or sub-shells of a shell (the third shell in the case of the elements from scandium to zinc, the fourth in the case of the elements from yttrium to cadmium, and the fifth in the case of the elements from lanthanum to mercury) after the first sub-shell of the next shell has been occupied (usually filled). Thus the electronic configurations of an awkward set of elements such as those in group III in Table 5.5 are as set out in Table 5.7. It will be noticed that boron, aluminium, gallium, and indium all have a 2, 1 pattern in their outer shells, whereas scandium and yttrium have two electrons in the first level of their outer shells and one electron in the third level of the shell within it. Thallium has the former pattern and lanthanum and actinium the latter. Those with the 1, 2 pattern are transition elements. Their electronic configurations are marked by a belated filling of the next-to-outermost shell. Thus the theory resolves the degree of resemblance Mendeleyev tried to cope with by talking of alternate series and later theorists by the device of 'short' and 'long' periods. The contemporary table places the transition elements in columns separate from those of the main set of the same valence. This is partly a consequence of a better knowledge of their properties. 'Row' relationships have been seen to be more striking than 'column' relationships in their case. But it is also partly a consequence of theoretical enlightenment, including a new way of grasping the facts. To this extent we have here an interesting example of the way in which a successful theory affects the form in which relevant facts are stated. Just as theories cannot afford to be completely a-factual so facts when significant are not a-theoretical.

6 COMMENTS

THE BASES OF EXPLANATION

Description, prediction and explanation

To describe some situation or object involves setting down its features and properties. Thus we may describe hydrogen as a gas at normal temperatures, as having an atomic mass of 1.008, a valence of 1, as being highly inflammable in oxygen and as having one proton in its nucleus and one electron in orbit around its nucleus. The first four items in this description of hydrogen have been derived from its observation whereas the fifth component is derived from theorizing aimed at a rationale for some observed properties of hydrogen and the other elements. Thus, though descriptions are usually thought of as being statements about the observed characteristics of situations and objects, it may be necessary to recognize hypothetical descriptions. Adams and Leverrier offered some hypothetical descriptions of Neptune before its observation and Mendeleyev offered quite detailed, though hypothetical, descriptions of scandium, gallium and germanium before these elements had been observed. Because it is thus possible to make hypothesized descriptions it has sometimes been said that explanations are only further descriptions. As a preliminary to the examination of deeper philosophical issues which such an assertion raises, some remarks will be made about description, prediction and explanation.

Briefly put, the conventional meanings of these three terms may be expressed as follows: (i) a description says what an object or a situation is, (ii) an explanation shows how, or even why, it is so or has become so, and (iii) a prediction is a foresaying about its occurrence, its features, etc. in advance of its being observed to occur or to have such and such features, etc. Granted that a description is not confined to those statements derived directly from observations, descriptions may be explanations and may be predictions. Bohr's 'descriptions' of the sub-atomic structure of the elements explains Mendeleyev's periodic table and Newton's 'descriptions' of the forces operating on particles of matter and of their modes of operation explain the planetary motions as conceived by Kepler. The important point, however, is not that Bohr's and Newton's theories are just further 'descriptions' but that they provide premises from which the descriptions derived from observation may be inferred. That they imply the facts of observation is a condition of their being an explanation of those facts. Mere implication may not be enough to constitute an explanation; all that is being claimed at this point is that it is a necessary feature of scientific explanation or at least of the type of explanation so far stressed.

There are two views of what constitutes an explanation, with no doubt some views involving an admixture of the two. The first stresses insight, intuition and even mystical revelation. It is illustrated by Kepler's appeal to the regular solids for an explanation of the spacing of the planets in the Solar System and to a less degree by Newlands' assimilation of the periodicity of the elements to the periodicity of the tonal scale. More markedly rationalist theories provide even stronger illustrations. As has been said above, the sciences have with the passage of time come increasingly to avoid explanations in this sense, leaving any task of the sort to metaphysics or to religion. This is part of what is meant when it is said that there is now seen to be no conflict between science and religion. The second view of explanation is the one which has been asserted several times above, namely that we have explained some phenomenon when we can show that it follows logically from premises which include some general rule, principle or law and which have not been shown to be untenable. That is, the explanatory premises should imply the facts to be explained and should not be contradicted by some facts other than those being explained. Some claim that an explanation need not involve a general rule, principle or law; instead they claim that any set of premises which enable the inference of the matter which is to be explained, the *explanandum*, would be regarded as providing an explanation. This issue cannot be taken up until some contrasts between rationalism and empiricism have been examined. Again, it may be said that premises including a general rule, principle or law provide an explanation whether it is tenable or not. It was said above that Eudoxos' and Ptolemy's theories provided an explanation of various celestial phenomena. They are no longer accepted because they are untenable and because they have other inadequacies, but it may still be fair to say that they do explain some phenomena. Of course, no one today says either provides *the* explanation of these phenomena—the complex of theories with which the names of Copernicus, Kepler, Newton, and Einstein are associated are deemed to provide the only explanation presently acceptable.

Prediction in science is a foresaying before the observation of the event but not necessarily before the occurrence of it. Neptune existed before Adams and Leverrier 'predicted' it, stellar parallaxes before the successors of Copernicus recognized them to be a consequence of his heliofocal theory and the elements scandium, gallium and germanium before Mendeleyev saw the need of them to fill in blanks in his proposed table. Apart from pure hunches that something or other may be the case, at least three kinds or modes of prediction in science may be recognized. These are illustrated first by Mendeleyev's anticipation of unobserved elements, and by the anticipation on the basis of Bode's law of a planet between Mars and Jupiter, second by Adams' and Leverrier's anticipation of a planet roughly corresponding to Neptune later observed by Galle and third by the implication of stellar parallaxes by the heliofocal theory.

Mendeleyev and Bode suspected a pattern among the observed events which was not fully supported by observation. If the pattern suspected were sound then some components of it not yet observed might be assumed in accordance with processes of interpolation and extrapolation. Adams and Leverrier, on the other hand, had a general theory, Newton's, which could account for the observed facts of Uranus' orbit only if another planet of such and such mass and orbit existed. They saw the need for an additional premise in deducing Uranus' orbit and were able to write it. The premise was about an observational matter and hence it was predictive because such a planet was later observed. Unlike the prediction of scandium because it was needed to fill in a gap in a suspected order or pattern of elements, and the prediction of Neptune because it was needed to provide the full set of premises to account for perturbations in the observed orbit of Uranus, the prediction of stellar parallaxes followed as an implication of assumptions made to explain other phenomena.

The first kind of prediction arises from the rounding out of a descriptive theory, the second from the provision of additional premises needed to enable an explanatory theory to explain known facts and the third as the implications of a theory formulated to explain other facts.

There are no certain recipes for theories

During the middle ages and early modern times in Western Europe, there was a marked tendency to seek the elucidation of everyday events by showing that they followed from some necessary or indubitable state of affairs. The Greeks had shown that some seemingly fortuitous facts were the logical consequences of 'self-evident' truths. Thus the theorem that the sum of the included angles of a triangle is equal to the sum of two right angles may be deduced from Euclid's axioms and definitions which seemed self-evident. Following a number of Islamic theologians, it was usual for medieval scholastics to accept that God was a necessary being, that is, one which must exist. Peter, Paul, Dobbin the horse, Stonehenge the monument and even the Earth need not in themselves exist. They are contingent upon God. In creating them or their makers, He left His marks on them. Thus an understanding of His nature confers some understanding of theirs, granted they be not completely accidental.

The words 'necessary' and 'contingent' have had shifting meanings, hence a particular commentator in order to pin the meanings down is likely to be dogmatic. It may help to consider the Latin origins of the two words. 'Necessary' comes from the Latin adjective *necessarius*, meaning unavoidable or inevitable and extended to mean indispensible, needful, pressing and urgent, and 'contingent' comes from the Latin verb *contingere*, meaning to touch together, to affect, to happen. In medieval usage their meanings were respectively 'inevitable' and 'dependent' or that which is inherently what it is and that which is what it is for some cause or reason outside its own nature. Modern usage is rather different. 'Necessary' has been narrowed somewhat to mean that which one would be involved in self-contradiction in denying and 'contingent' has often come to be taken to be what the scholastics following Aristotle would have called accidental or fortuitous.

It happens to be the case that the star-like planets under telescopic observation have visible discs and that blood under microscopic observation is seen to consist of a clear liquid bearing red and white corpuscles. There is no inherent reason that these situations should be so. It could have been that the fixed stars but not the star-like planets presented in the telescope were visible discs and that blood under the microscope appeared, as it does to the naked eye, a continuous substance. Being a planet and having a telescopically visible disc are matters which might or might not have come together. This is a contingent fact in that it does not have to be as we have found it; it could have been otherwise had nature been different from what it is. It is important to recognize that though contingent in this sense, it is not literally accidental or random. Contingent facts may be regular and stable; it is merely that they could have been different, perhaps if other circumstances had been otherwise. For instance had the fixed stars been near and the wanderers far away the former would have had telescopically visible discs and the latter not. Hence 'contingent' is still deemed by some thinkers to carry not only the meaning 'happening together' but also the meaning 'dependent upon' something else. However, it is taken by some to mean merely that some situation could have been other than it is.

Contrasted with contingent truths are necessary truths, relationships or states of affairs which in themselves could not have been other than they are. If one can find necessary truths which imply contingent facts, then one has found unshakeable rationales for what would otherwise have been brute facts. Some thinkers have held that such necessary truths may be attained intuitively, others that they are known through some process of insight and still others that they are given in revelation either by an unquestionable authority or through some mystical process. A feature of modern thought has been to call such notions into question. For instance, Francis Bacon in his *Novum Organon Scientiarum* (1620) proposed that general laws or principles could be induced from the observed facts. Had he been right, he would have successfully turned on its head the medieval view that the particular observed facts are established and secured by being deduced from general truths of an axiomatic kind. Ever since he proposed this new instrument of science there have been optimists who believed they could show its logical validity.

Deduction is the unfolding of what follows logically from premises. In other than very simple inferences, such premises must include general or strictly speaking universal propositions. However, except in a very limited case, what we are able to observe is particular. That is, we observe one or at most a few instances at a time, and the number of observations we can make is finite. Thus, except where what we want to talk about consists of a small finite class, we can have observed only some of the instances. We say that we have observed that whenever an electric current passes through a mixture of hydrogen and oxygen, water is formed. This may be true of all the instances we have observed: we may assume, but we cannot validly infer, that it is true of all the instances we have not. The breakdown of Bode's law is a clear demonstration of the frailty, indeed the rashness, of inducing general statements from the observation of particular cases. Sometimes the generalization breaks down almost as soon as further observations are made; sometimes it holds up no matter how many additional observations are made. Unfortunately no one has come up with an unfailing recipe for developing viable generalizations by induction from particular observed facts.

The efforts to state and to understand the spacing of the planets and to state and to understand the periodicity of the elements provide an instructive contrast. Both began with a recognition of a hint of some pattern. Just as Kepler's fitting of the planetary orbits within spaces defined by the regular solids had a mystical aura so too did Newlands' parallel of the series of elements with the tonal octave. Newlands' parallel was not so good as not to be true—it merely added a little plausibility. Kepler's regular solid fitting was nearer to that state. Kepler's was a completed or closed schema as there were no more regular solids. This proved a fatal weakness when additional planets were discovered. Newlands' schema on the other hand would have been strengthened by the discovery of additional elements provided they possessed some rather than other properties. The discovery of scandium was 'right' for him but not the discovery of the inert gases for they required nones rather than octaves and thus destroyed the mystical parallel.

These early attempts to find a pattern and some sort of rationale were followed by the more realistic attempts by Bode and by Mendeleyev respectively. Mendeleyev had better luck or a better hunch than Bode and as a result the two stories end quite differently. Both proposals needed certain additional discoveries and could accommodate still other discoveries. Ceres, Pallas, Juno and Vesta when observed filled the gap for Bode between Mars and Jupiter, and Uranus though not needed by the schema did not discommode it. Neptune and Pluto quite decisively upset it.

Mendeleyev gave himself more flexibility than Newlands or than Bode when he introduced a double classification—by order of atomic weight in the rows and by valence in the columns. This allowed him to set series of different lengths and enabled him to leave gaps in the series with more justification. It was such flexibility conferred by complexity that contributed to the long survival of Ptolemy's theory. The advantages it brings are thus counterbalanced by the increased risk of being saddled for too long with an erroneous theory. However, it was the greater complexity of Mendeleyev's theory which enabled him to predict many more properties of scandium, germanium and gallium than Bode could have predicted for the planetoids or for Uranus. Bode's law enabled the prediction of a mean distance from the Sun and nothing more; Mendeleyev's table enabled the prediction of atomic weight, valence, and a number of other properties including the readiest method of detection.

Mendeleyev was lucky in the errors in some of the information available to him. For example, he might have been put out of his stride had he known that tellurium was really heavier than iodine, and not lighter as his evidence suggested. Later it became possible to abandon his basis for placing the elements in an order, namely the order of their atomic weights. Consequently when the true order of atomic weight of iodine and tellurium was established it could no longer make difficulties for the table.

In respect of the successive adjustments to various details, the periodic table was not at first basically different from the Ptolemaic theory or the original Copernican theory. In all three, the basic notions as well as the details could be stretched or compressed enough to fit the facts. All three were to a greater or less degree *ad hoc*. Just as Newton provided a rationale for the Keplerian version of the Copernican theory, so Bohr showed why Mendeleyev's table obtained. What had in both cases been brute facts, in so far as they were facts, became intelligible or explicable facts. Though Newton and Bohr may prove to be wrong in their suppositions, the very situation in which their theories account for a wide range of observed facts and have enabled the anticipation by inference of facts not previously observed increases the probability of their being correct, granted that better alternatives are not available. Brute 'facts' for which no rationale has been found are more likely to suffer later modifications or even rejection than are 'facts' for which an explanation in some wide ranging terms has been provided. It is worth noting as an additional point that Newton had the explanation of Kepler's laws in mind when he formulated his theory, whereas Bohr had in mind some problems quite different from the explanation of the periodic table. This, however, makes no difference to the cogency of the explanation, though it may affect our assessment of the theories' probabilities.

There are no recipes either for the discovery of patterns and orders or for the invention of adequate rationales for those patterns which have been discovered. Some orders based on partial sets of observed facts prove to fit the facts that are subsequently observed, some do not. Some theories turn out to explain facts not known at the time of their invention, some are falsified by new facts. Except in those rare, if ever realized, cases in which all the facts are already available, any proposed order or any proposed explanation can be upset by facts which are subsequently garnered or replaced by more inclusive orders or more illuminating explanations that are produced. All that can be done is to test the hypothesis, after its proposal, by means of facts other than those guiding its formulation. As we shall see it is important that the hypothesis be open to falsification; that is, there should be a way for the facts to turn out which is contrary to the hypothesis. If they turn out to be as the hypothesis implies, then the hypothesis has that much more support from the evidence relevant to it. If they turn out to be contrary to what the hypothesis implies, then the hypothesis is falsified.

Two main views of theory and fact

Two main philosophical attitudes adopted in modern times to a number of logically distinct but historically related issues bear on the relations between theory and fact in science. They are labelled rationalism and empiricism, words which originate in the Latin *rationalis*, reasonable, depending on reason, and the Greek *empeiros*, experience, proved by experience. They are complex sets of views with many variants. Few, if any, philosophers have been pure and complete in their adoption of one or the other. Hence what follows is not to be thought of as an exposition of the views of any particular philosophers. Rather a composite picture of each of the two philosophies is being presented. More systematic historical treatises should be turned to for an account of the views of individual philosophers.

A principal issue upon which rationalists and empiricists divide is the basis upon which we should seek a justification for our beliefs about nature. The term 'nature' is used here in the widest sense to include the physical universe from atoms to galaxies, the organic realm from

complex organic molecules to multicellular organisms and the whole
of human nature from the traits and habits of individuals to the structure of human society and the products and media of human intercourse
such as art, technology, law, religion, and science. Rationalists and
empiricists are not in dispute about a wide range of contingent or
brute facts such as that hydrogen has a valence of 1 and oxygen a
valence of 2, that Mars has a more eccentric orbit than Venus, that
venous blood is dark red whereas arterial blood is bright red, and that
when tall and dwarf garden peas are crossed the hybrid offspring are
tall though some of the tall hybrids' offspring are dwarf. The rationalist
considers such observed truths to be so ephemeral, so isolated, so
a-rational and so much a mere conjunction of properties and events
that he seeks a justification for his belief in them in some more secure,
abiding rational principles. The empiricist looks to the experience
of these situations themselves for his justification for believing in them.
The empiricist may, as the rationalist does, look outside these contingent truths for an explanation of them. The rationalist regards the
explanatory theories of science as a step, if not the whole journey,
towards the basic necessary truths, whereas the empiricist considers
that no matter how the theories enlighten the facts they must be tested
against the facts. Their contrasting views on this issue not only render
the use of labels derived from *rationalis*, depending on reason, and
empeiros, proved by experience, appropriate, but are central in the
aspect of the philosophy of science with which we are concerned. Hence
we shall examine them more fully below.

A second issue on which rationalists and empiricists tend to divide
is the way in which we attain knowledge. Arriving at a belief and
justifying it are different matters. Mendeleyev could have thought of
his periodic table in a dream, he could have had it come to him as a
flash of insight while bathing or he could have worked tortuously and
slowly towards it by writing the elements down in various patterns on
innumerable sheets of paper. But how he arrived at it and how he
could properly convince both himself and others that he had arrived
at something of worth are different matters. Some rationalists have
claimed, as Descartes did, that the basic truths are inherent in the
human mind, that is, that they are inborn ideas. Others have suggested
an intuitive access to basic truths as a complement to the sensory access
to contingent truths. Still others have turned to the notion of self-
evident truths; more recently those in this group have argued that

some truths can be recognized to be necessarily true as a denial of them would involve one in self-contradiction. A distinction is commonly made between *a priori* and *a posteriori* truths, those statements whose truth may be discovered *prior to* and those whose truth may be discovered only *after* the observation of relevant situations, or better, those statements whose truth can be shown independently of any observation of relevant situations and those whose truth can be shown directly or indicated indirectly by observation. The empiricist, being chary of anything not proved by experience, denies or minimizes innate ideas, intuitive or self-evident truths and *a priori* knowledge. In place of each he substitutes knowledge by experience; some strict empiricists wish to confine us to the knowledge gained by experience whereas more liberal empiricists also allow knowledge either constructed from or tested by knowledge gained by experience.

This issue of inherent versus acquired knowledge remains in one aspect a problem for psychology, which interestingly treats it as an empirical question. For instance, are properties such as red, square and dual inherent in visual perception or do they have to be learned? Though this is an interesting question, it is not important for us here. The issue also has some bearing on the relation of logic and mathematics (the formal sciences) to the observational sciences such as physics, biology, psychology and anthropology. We shall look briefly at this relation later.

A third issue upon which rationalists and empiricists tend to divide concerns levels of reality involving a distinction between the superficial and the essential. In everyday thought or commonsense, which is often the unquestioned, popular residue of an earlier and contentious philosophy, we often make a distinction between the superficial and transient features of some occurrence and its essential, enduring character. This distinction is also to be found in science. Thus the essential, enduring water is preserved even though the substance is manifested in the superficial, transient properties of gaseous steam, liquid water or solid ice. More profoundly it is expressed in the idea that oxygen is oxygen whether free as O_2 or as O_3 (ozone) or whether in combination as in water (H_2O), in sulphuric acid (H_2SO_4) or in the complex proteins which are so important in the structure and economy of living organisms. The rationalist is prone to think that these are essential natures which are comprehensible only by reason but which manifest themselves now in this form and now in that, these forms being all that sensory perception may apprehend. The empiricist shies away from such a notion because he wonders how he can get to the essential natures of things except through his sensory perceptions and why he should not consider things to consist, at least in part, of what he perceives them to be. In general, scientists are reluctant to line up with the more thoroughgoing empiricists on this issue. The notion of an underlying, essential nature has proved attractive to them.

A fourth issue, which has frequently led rationalists and empiricists, especially the latter, to cross the floor of the House in order to join the Opposition, concerns elements in the sense of irreducible atoms of a simple type. Because rationalists in the seventeenth, eighteenth and nineteenth centuries were often holists, the empiricists of this period were most often elementarists. Yet elementarism is by temper a rationalist doctrine. It subscribes to the proposition that things as they occur are reducible to a limited number of irreducibles. Such a notion of getting to a bedrock is congenial to rationalism rather than to empiricism. Rationalism has usually looked for the relatively simple. This may be the One, the simplest even if the most complex, which is espoused in holistic views (the word 'universe' betokens a subscription to the most embracing holism of all—everything is one) or it may be the basic few such as the early Greek elements of Earth, Water, Air, and Fire or, more thoroughgoing, the early Greek properties of Hot, Cold, Dry, and Moist. Mendeleyev's table of elements has too numerous entries and too many anomalies for it to be fully congenial to a true rationalist.

A fifth issue, the views on which derive in part from decisions taken on the issue of the essential versus the manifest and on the issue of the whole versus the parts (elements), centres on reducibility. It is not insignificant that Newton saw the motions of planetary bodies to be reducible to or deducible from the laws of motion of particles of matter, that the pattern of properties among the chemical elements was seen, with Bohr's aid, to be reducible to the sub-atomic structure of these elements and that, as we shall see in the next Case Study, the key that may unlock the mystery of heredity involves the complex structure of the organic chemical molecule DNA. To put the matter simply, it would seem that a biological puzzle such as heredity is reducible to chemistry and a chemical puzzle such as the facts of the periodic table is reducible to sub-atomic physics. The general flavour of this is

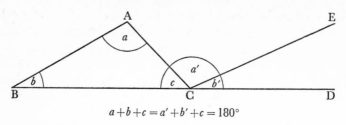

$$a+b+c = a'+b'+c = 180°$$

Fig. 6.1 Construction from which it can be 'seen' ('intuitively') that the included angles a, b and c, sum to two right angles, $a' + b' + c$.

rationalistic. However, it is paradoxically the case that empiricists have more often subscribed to reductionism than have rationalists and that rationalists have more often opposed reductionism by propounding some holistic notions in which complexes are endowed with properties inexplicable in terms of anything simpler. The idea of a hierarchy of sciences and the consequential notion of explanation by reduction will be examined more fully later.

Formal systems and formal sciences

The ancient Egyptians, Babylonians and Greeks discovered in an empirical way many relationships which were later incorporated in deductive mathematical systems such as Euclid's geometry. Pythagoras' theorem that the square erected on the hypotenuse of a right-angled triangle is equal in area to the sum of the squares erected on the adjacent sides is perhaps the most widely mentioned example. Another is that the sum of the included angles of a triangle is equal to two right angles. We learned at school that both of these theorems could be proved by argument from Euclid's axioms and definitions. In the case of the second we possibly felt that such a long run up was unnecessary, as one can readily 'see', intuitively as it is often put, that it must be correct. Consider the figure usually used to give a little specificity to Euclid's abstract argument. The triangle ABC in Fig. 6.1 has its base BC extended to D; another line CE is drawn from C parallel to the side BA. Angle b' clearly looks equal to angle b (and application of a protractor will provide corroboration) and angle a' to angle a. The included angles are a, b and c. Their sum is manifestly equal to the sum of a', b' and c, and the sum of the latter is manifestly 180° or two right angles (again a protractor will provide corroboration).

Though inspection of any particular triangle will convince us that the relationship holds for it, we may be left with the nagging doubt that there may be some triangle which we have not considered for which it does not hold. Bode's law held not only for Mercury, Venus, Earth, Mars and, after a gap, Jupiter. It held for the first observed planetoids and for Uranus, but then it broke down with Neptune. What assurance have we that no Neptune of a triangle is waiting like a villain in the wings to interrupt the happy ending to the present play? This nagging doubt is always likely to rise as we make the inductive leap from the broken light of our incomplete facts into the complete darkness of what we have not yet observed. What makes Euclid so important is that he provides a safe bridge across the chasm, or what seems a safe bridge. He shows that, no matter what the dimensions and the angles of the triangle, the theorem follows as a logical necessity from his axioms and definitions. The latter constitute the anchorage for a logical bridge to the general truth manifested in the particular cases.

Amongst the axioms relevant to the theorem about the included angles is one which may be referred to as the parallel lines axiom. It is that one, and only one, straight line may be drawn through a point external to a second straight line which does not intersect that line no matter how far the two lines be extended. Over the centuries mathematicians have not been as convinced about the certainty of this axiom as they were of the others. It did not appear so evidently true. Numerous attempts were made to deduce it from the other axioms or from them when supplemented by 'axioms' more acceptable than it. Lobachevski and Bolyai, working independently in the years around 1830, substituted an axiom asserting that more than one straight line could be drawn through a given point so as not to intersect a given straight line to which the point was external. Instead of producing confusion and self-contradiction, they generated a geometry (or more exactly two slightly variant forms of a geometry) alternative to Euclid's. Some of their theorems were the same as his and some were different. For example, Lobachevski has a theorem to the effect that the included angles of a triangle sum to less than two right angles. Speaking pictorially Kemeny has said that surfaces, though plane in the sense of being two-dimensional, 'sag' in Lobachevskian geometry as Fig. 6.2 depicts.

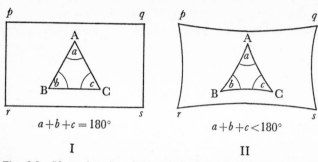

Fig. 6.2 If we let the plane *pqrs* in I 'sag' as in II, then the Euclidian triangle ABC in I becomes a Lobachevskian triangle in II.

It usually comes as a surprise to the non-mathematician that such a geometry should not generate self-contradictory theorems when it generates theorems, like its included angle theorem, which are 'obviously' contrary to nature. It comes as a further shock that evidence from spaces pegged out by remote galaxies suggests that the Lobachevskian theorem is a better fit to nature than the Euclidian theorem. This, however, is not the main point to be made. The more important thesis is that alternative internally consistent geometries can be developed independently of the way spatial properties and relations occur in nature. This strongly suggests that a geometry need not refer to nature at all in order to be worthy of respect as a completely abstract or formal system. Further consideration suggests that the axioms and definitions in such a system may be arbitrary, provided they are self-consistent. Geometry, a word derived from the Greek *ge*, the earth, the land, and *metrein*, to measure, involves many basic terms which may be given an interpretation having special relevance to land surveying. A point may be thought of as the centre of a surveyor's peg, a line as a surveyor's chain stretched between two pegs, an angle as the degree of separation in direction of two chains stretched from a common peg to two separated pegs, and a figure as the shape of any area bounded by chains stretched between three or more separated pegs or traced by means of a continuous chain looped over one or more pegs. These terms need not, however, be given these empirical meanings. They need not be regarded as having any referents whatsoever in nature for an internally consistent geometry to be generated around them. They could be empirically unidentified terms *a*, *b*, *c*, etc. entering into empirically unspecified relations, *r*, *s*, *t*, etc., and still the set of postulates (the propositions demanded or taken for granted) could be unfolded into logically consequential and logically consistent theorems, which also would lack empirical referents. Such a system of postulates and definitions generating theorems through rules of inference constitutes an abstract formal system. It may be left without empirical interpretation just as a work of abstract graphic art may be left without a pictorial interpretation. In what sense such a formal system remaining uninterpreted constitutes knowledge is something of a puzzle. It is hardly knowledge of nature and, as it can not appropriately be said to be true or to be false, it is hardly knowledge of anything else. It is rather like an intellectual five-finger exercise: it is well or ill executed and that is about all one can say of it.

When it is interpreted, that is, when its terms are given empirical referents, the situation is transformed. The formal system when interpreted enables us to draw conclusions about the material in terms of which it has been interpreted. In some factual cases, any one of several formal systems applies equally well. Within the cramped space available on the Earth's surface it is not possible for us to establish whether the sum of the included angles is 180° or a little less or a little more: 180° seems a good average. The straight lines are hardly long enough to reveal the effect of a slight 'sag' or a slight 'bulge' in the planes. We need to examine a much more extensive space such as that marked out by galaxies almost unbelievably remote from one another in order to find evidence that Lobachevskian geometry with its sagging planes gives a better fit to nature than does Euclidian geometry with its 'flat' planes or Riemannian geometry with its 'bulging' planes.

The issues raised here are difficult and controversial and it is not possible to pursue them as far as they deserve. They have been taken far enough, however, to provide a better insight into the contention that at least some theories are to be regarded not as conjectures about what obtains in nature but as calculating devices for use in dealing with natural phenomena or as effective ways of thinking about and drawing inferences about what has been or may be observed.

Four views about theories

Theories may be more or less distant from the facts relevant to them. The assumption that the stars complete below the horizon the circles whose arcs they are seen to draw during the evening above the horizon is not 'far' from the observed facts. It is indeed merely an extrapolation from them. Kepler's theory of planetary motion—elliptical orbits and so on—is not just an extrapolation from the successive observed positions of the planets in the sky. They are allegations of motions which are not observed in part as the stellar circles are; they were invented by Kepler and they are different in kind from the statements of the apparent motions of the planets which they imply. The assumption of the complete stellar circle enables the inference of the observed paths of the stars. If a star moves only on a given circle, then it necessarily follows that the star moves on any arc of that circle, part of which it traces above the horizon. This is a very simple inference, being in the class that is called immediate; an example is provided by the argument: 'If all dogs are vertebrates, then this dog is a vertebrate'. To move from Kepler's laws of planetary motion taken in conjunction with some facts about periods or distances to the observed path of a planet in the Earth's sky requires many inferential steps and in this sense the theory is more distant from the facts than 'all dogs are vertebrates' is from 'this dog is a vertebrate'. Newton's dynamic theory in addition to being more abstract is still more distant from the facts which may be inferred from it.

Similarly, though the recognition of the halogens as a family of elements may involve going a little beyond the facts, it does not go as far beyond them as does the assumption that patterns of family relationships will be brought out by arranging the elements in order of atomic weight and in columns in terms of valence. Still more distant from the facts concerning the properties of the elements are Bohr's assumptions about nuclei and electron shells.

In general the more distant the theory from the relevant facts the more it appears to provide an explanation of them and the wider the range of facts it explains. There seems to be hardly any explanation of the arcs constituting the visible paths of the stars above the horizon in the contention that they must be so because the stars' full paths are circles. There does seem something more of an explanation of the apparent planetary paths with their speeding up, slowing down, standing still and retrogressing in our deduction of them from Kepler's three laws taken in conjunction with the periods of the Earth's and the other planets' revolutions around the Sun. This is a genuine explanation even though it may be said to be at a low level in that Kepler's laws themselves have a descriptive quality and obviously call for an explanation of themselves. In contrasting Kepler's laws as descriptive with Newton's as explanatory, philosophers of science have meant to emphasize that the explanation given by Newton's theory is more profound and more telling. Of course, one may still ask for an explanation of Newton's theory, for it is not necessarily the ultimate.

One extreme view of scientific theory, a rationalist view, is that as the theory goes 'deeper' or more distant from the 'surface' facts, it approaches more closely to the necessary truths which being necessary are not in any need of explanation themselves. The rationalist evaluates a theory both in terms of its being ultimate or of its approach to the ultimate, and in terms of its being in no need of explanation itself because it is a necessary truth or of its having a fairly plausible mien suggesting that even though it could be otherwise it is hardly likely to be.

The question whether there are any necessary truths other than those of a trivial kind ('Bachelors are unmarried males') or those involving assumptions we don't care to question ('If the Defence Minister were in Canberra on that day, he could not have been cruising in the Caribbean') will not be examined here. The relevant question for us is whether any propositions for which the status of necessary truth may be claimed have relevance in explaining any observed facts of interest in science. It has sometimes been suggested that chemistry is an extension of physics and our third case study seems to support that notion. Likewise physics is said to be an extension of mathematics which in turn is said to be an extension of logic. In our next case study we will encounter an instance where biology seems to be an extension of chemistry. It would not be difficult to find cases where psychological phenomena seem dependent on, and perhaps an extension of, more basic biological events. Some reductionists, however, consider that the process of reducing one science to another must stop with physics. Physics, chemistry, biochemistry, biology, psychology, and sociology are deemed to be factual or empirical sciences whereas logic and mathematics are deemed to be purely formal sciences. That is, Newlands' table has been rejected and Mendeleyev's has been accepted after modification because the former did not and the latter did fit the facts. Euclid's geometry, Lobachevski's geometry and several other geometries are all 'accepted' not because they all fit the facts but because they are all internally consistent. If physics is an extension of mathematics and if the basis for deciding between alternative physical theories is the set of relevant facts, then some geometries may have to be rejected on the grounds that they do not fit the facts. Setting aside for the moment the question of whether or not biology is reducible to biochemistry, biochemistry to chemistry and chemistry to physics, it may be said with plausibility that the way logic and mathematics are involved in physics, chemistry, biochemistry, etc. is different from the way physics may be involved in chemistry, physics and chemistry in biochemistry, and so on. Logic and mathematics are involved in the other sciences in their provision of modes of analysis and inference. On the other hand Bohr's physical theory is involved in the chemistry of Mendeleyev's table of the elements through its provision of bases or premises from which the pattern of chemical phenomena may be deduced. Whereas logic and mathematics are concerned with form upon which analysis, argument and inference depend, the empirical sciences such as physics, chemistry and biochemistry are concerned with content, that is, what is analysed, what is argued about and what inferences are drawn from or drawn about.

As no theories or laws in the empirical sciences other than the laws of inference have any supportable claim to being necessary, the rationalist view of theory must be rejected. Nevertheless something of what it maintains is worthy of consideration. Though the more 'general' and more 'abstract' theories may not be principles quite in the rationalist sense of that which not only does and can only come first, the *explicans* which needs no explanation, they do seem different from other generalizations of low level explanatory value, generalizations which are close to the facts.

On the other extreme from the rationalists are the ultra-positivists or strict empiricists who are distrustful of theorizing. The more speculative theorizing is dismissed by them as an indulgence in metaphysics. The embarrassment in which physics found itself at the end of the nineteenth century through its assumption of an aether as a medium for the transmission of light is cited as a warning example. Talk of explanatory laws and concepts implying a contrast between explanation and description is deplored. Explanation, say the ultra-positivists, is only more detailed description. Theorizing should confine itself to the analysis and ordering of what has been observed and where additional data are assumed they should be of an observable kind. Suppositions about additional planets or additional elements required to complete some pattern are in order provided these items are observable and provided that the ultimate test is either their observation or their non-observation in circumstances which should permit their observation if they indeed exist. Laws such as Newton's law of gravitation are deemed to be ingenious generalizations of the facts. The pull of the Earth on the Moon as revealed in the Moon's deflection from a tangent to its orbit is given by Newton's formula

$$F = G \, m_1 \, m_2 \, / \, r^2$$

The law expressed in this formula is held not to be something other than the facts, it is something manifested in all the relevant facts if only one has the wit to see it.

Two other empiricist views of theory recognize that something outside the facts are introduced in theories. Neither view claims the special status for this material that the rationalist does. It is not held to be something in a secure realm of necessary truths or at a 'more real' level of basic natures. Osiander and Bellarmino were early proponents of one of these views and Galileo of the other. Osiander and Bellarmino held that the Copernican heliofocal theory of the planetary system might be validly regarded as a mode of thinking about the system which enabled easy and effective calculation of the apparent motions of its members; they did not wish to regard the theory as a conjecture about what was the case in the system. The instrumentalist view of Osiander and Bellarmino has been worked out more fully and put more convincingly in the twentieth century by a group of philosophers of science often labelled logical positivists or scientific empiricists who may also be called instrumentalists, as they held theories to be instruments enabling inferences from one set of facts to another. The twentieth-century upholders of a view like Galileo's are usually called realists, because they claim that theories are conjectures about what exists in reality.

The three types of empiricism briefly described above are not exhaustive; they are, as it were, the extreme points, the apices, of a triangle of various views. All empiricist views regard the observed facts as the basis for what one believes and adopts. Strict empiricism follows this policy in the most literal way. It is willing to venture only where the facts provide direct substantial support. Realism and instrumentalism are ready to take greater risks in skating on what may be thin ice, though both demand some manifest ice. They differ in the way in which they offset the added risks. The realist says that he is relying on additional ice about which he has made guesses rather than observations; he recognizes that he is in for a ducking if in fact the ice is not thick enough to support him. The instrumentalist says that what he has added is skating technique and he expects to get by not because the additional ice is there in fact but because his skating as though the additional ice were there is clever enough to get him across to the other bank.

A decision for a particular form of empiricism depends mainly upon the logical relationships which exist between fact and theory in science. Strict empiricism counts on little more support than the method of induction provides, whereas the other two, having little confidence in the support that induction offers, turn to what may be called the hypothetico-deductive method.

The hypothetico-deductive method

One may think of three ways of proceeding in respect of universal propositions and particular propositions where the former explain, order or predict the latter and where only the latter are directly derived from observation. First, universal propositions may be indubitable principles which are attained by some process of intuition or of insight or of revelation and which constitute the premises from which the particular propositions describing observed events may be deduced. This may be called the deductive method. Assuming that the theorems in the various branches of mathematics have an inherent factual reference, and not merely an additional factual application, a case can be made for the deductive method through an examination of mathematics. Second, universal propositions may be derived in some valid way from the particular propositions describing the observed events. There is a sound procedure whereby one observes every member of a class and reports anything they have been seen to have in common. Thus every planet in the modern sense observed from the Earth (this excludes the Sun and the Moon which the Ancients called planets, and the Earth which we call a planet) has been observed to have at least one phase of retrograde motion in each circuit of the sky. Though the observations have been of particular events one may justifiably summarize the knowledge derived from them in this generalization. We cannot, of course, on the basis of these observations speak with logical justification of other than observed circuits of these planets or of the circuits of other than the observed planets. Provided we restrict the reference of the generalization to the instances observed, the step from the particular cases to the general statement is induction by complete enumeration. Once the generalization embraces other than observed cases we are resorting to induction by incomplete enumeration for which there seems to be a commonsense justification. Unfortunately it lacks either logical or factual justification. The general statement about the circular paths of the fixed stars in their journey above and below the horizon is one that few would wish to question. The observation of the paths above the horizon constituting arcs of these circles, however, does no more than make the general statement plausible; the observations do not prove the generalization.

The more important universal propositions of science are not strictly generalizations of a number of observed particular cases. Tycho Brahe, Kepler or any other person on earth has never observed Mars or any other planet making elliptical orbits around the Sun with changes of pace such that equal areas are swept out in equal times by vectors from the Sun to the planet. These are notions different in kind and not just in 'number' from what has been observed. Kepler imagined his laws and then showed that they fitted the observed positions of Mars in the sky at given times. The facts can be inferred reasonably accurately from these laws. To say that the facts suggested the laws is to underestimate the creative contributions of Kepler. His hypotheses were formed with the observed facts of Mars' path in the sky in mind, just as Eudoxos' and Ptolemy's theories were formulated with some of the same facts in mind. Though the theory fits the facts, though they logically follow from it, it does not follow logically from them. It is not formally required by them and it is something additional to them.

The theory is not, however, logically independent of or unrelated to the facts upon which it bears. While the facts do not imply or logically require it, it, in conjunction with some other considerations, implies or logically requires the facts. Were the facts other than they are, they could show the theory (or the other considerations or both) to be false. The facts being as they are, the theory (in conjunction with the other propositions actually considered) is at least tenable: it may or may not be true, but it is not contradicted by the known facts.

These comments draw attention to three distinctive phases in the hypothetico-deductive method in the empirical sciences. The first consists in the formulation of an hypothesis—it does not matter how, whether cautiously or carelessly, whether painstakingly as Kepler is said to have framed his elliptical law or in a flash as Newton is said to have comprehended gravitation when he saw the apple fall. The second consists in unfolding the implications of the hypothesis, when taken in conjunction with other considerations (either based on observations or other assumptions), for some observable events. In the third, the theory is declared tenable when the implied events are corroborated by observation or declared false when the implied facts are shown by observation to be false. Sometimes for technical reasons the relevant observations prove not to be decisive. The early attempts to establish stellar parallaxes provide an example of such indecisive evidence. In the third phase it is important that the hypothesis be tested through situations other than those which prompted its formulation. Such additional testing goes beyond checking that the hypothesis meets the requirements its inventor had in mind. When it is confirmed in these unanticipated situations its probability is markedly enhanced. Further, the range of its explanatory and predictive power is demonstrated to be wider than may have been originally thought: for reasons of economy of assumption, if for no other, theories of wide range are more highly esteemed than those of narrow application.

The three phases just summarized do not always, as a matter of historical fact, occur in the order listed. Sometimes the observations come first, sometimes observations are prompted by hypotheses without the strict unfolding of implications of the second phase. But this does affect the formal structure or logical properties of the method.

Part Four

INHERITANCE AND THE
GENE THEORY

7 *CASE STUDY*

The two preceding case studies have presented an initial discovery of a pattern in nature and a subsequent explanation of it. The main features of the pattern of apparent celestial motions were discovered in Babylonian times. Details had to be added and refinements made over the centuries but the basic facts, the main outlines of the picture, were not altered. Attempts at adequate analysis and explanation failed until the sixteenth and seventeenth centuries AD when in a little more than a century Copernicus, Kepler, and Newton forged the key which unlocked the problem. By contrast, the discovery of the pattern in the properties of the elements was late and comparatively quick, and it was followed in fairly brief time by an explanation. The case study to be presented now is different from the two preceding it. That there is some regularity in the inheritance of characteristics by offspring from their parents was recognized in antiquity. That each species begets only its own kind and that within a species individuals tend to beget their like are generalizations of long standing with good evidential support. Further, though the pattern of inheritance, much less its rationale, was not grasped, plant and animal breeders down the ages understood enough to be able by selective breeding to strengthen the occurrence of traits they wanted and to reduce the occurrence of traits they did not want. Unfortunately their traditional lore was rife with misunderstandings and superstitions which though incredible today were then given as secure a place as that occupied by the scattered genuine data available.

Mendel's contributions

We are indebted to Gregor Mendel for our first clear insight into part of the pattern of inheritance, a simple part in a tremendously variegated design. He reported in 1865 an impressively clever, patiently methodical and possibly lucky series of experiments on the garden pea. He chose

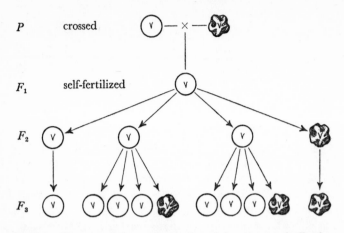

Fig. 7.1 Diagrammatic representation of the pattern of offspring of self-fertilized hybrids (F_1) of round and wrinkled seeded parent stock. The pure and hybrid round seeded F_2 generation can be distinguished only by cramming their offspring in the F_3 generation.

this plant for several reasons. First, it had easily discernible features such as seed shape, pod shape and stem length, each of which varied markedly among the several varieties of the plant. Second, there were several well-known varieties each of which, when allowed to self-fertilize, bred true. Third, though the plant normally self-fertilized without being appreciably affected by air-borne or insect-borne pollination, it could be easily artificially fertilized in a controlled way. To this clever choice of a plant, he added two clever choices of method; first, a resort to a large-scale investigation which minimized what we now call sampling errors and second, a resort to quantitative data. The luck, if it were such, was the choice of seven traits—shape of the ripe seed, colour of the cotyledons, colour of the seed coat, shape of the ripe pod, colour of the unripe pod, axial or terminal position of the flowers on the stem and the length of the stem—in an organism which we now know has seven pairs of chromosomes in the nuclei of its cells, the genetic determinant of each of these seven traits being located in a different chromosome pair.

He began his experiments with plants which he had shown over several generations bred true if allowed to self-fertilize. He then artificially crossed pairs of plants differing in respect of one or more of the traits. After this initial hybrid fertilization, he allowed the subsequent generations to self-fertilize. Let us consider the case of crossing plants with round seeds and plants with wrinkled seeds. He made sixty fertilizations on 15 plants. All seeds, constituting the first and hybrid generation (F_1) resulting from these fertilizations, were round. When 253 plants grown from these seeds were allowed to self-fertilize, they produced 7,324 seeds of which 5,474 were round and 1,850 wrinkled. The proportion of round to wrinkled seeds in the second generation (F_2) approximates 3 : 1. Though only the round trait appeared in the first generation, both the round and the wrinkled traits appeared in the second generation. Mendel, therefore, regarded roundness of seeds as a dominant trait and wrinkledness as a recessive trait. Plants in the third generation grown from the wrinkled F_2 seeds when allowed to self-fertilize produced only wrinkled seeds (in the F_3 generation). Of 565 plants grown from F_2 round seeds, 193 produced, after self-fertilization, round seeds only and 372 produced both round seeds and wrinkled seeds in a proportion approximating 3 : 1. The 193 F_2 plants behaved like their pure round-seeded progenitors of two generations back, just as the plants germinated from F_2 wrinkled seeds behaved like their pure wrinkled-seeded progenitors of two generations back. On the other hand the 372 F_2 plants germinated from round (F_2) seeds behaved like their hybrid F_1 parents. Thus in terms of the F_3 seeds produced by the F_2 offspring of the hybrid F_1 plants, three F_2 types may be recognized, namely (i) round seeds developing into plants yielding only round seeds, i.e. a pure type, (ii) round seeds developing into plants yielding both round and wrinkled seeds in a proportion of 3 : 1, i.e. a hybrid type and (iii) wrinkled seeds developing into plants yielding only wrinkled seeds, i.e. a pure type. The three types in the F_2 generation were found by Mendel to be in the proportions 1 : 2 : 1 approximately. These facts are illustrated in Fig. 7.1. A similar pattern was found with the other six pairs of traits. Yellow cotyledons (the two 'halves' of the pea seed which on its germination are transformed into the primary pair of leaves) were dominant to green, coloured seed coats to white, inflated pods to constricted, green pods to yellow, axial flower placement to terminal placement and long stems to short. These patterns occurred irrespective of which parent provided the pollen and which the ovules in the initial cross-fertilization. Only shape of seeds and colour of cotyledons are affected in the seeds resulting from cross-fertilization, the other traits appear

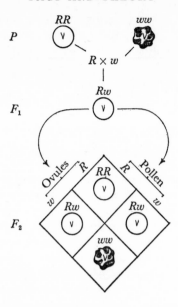

Fig. 7.2 Mendel's explanation of the pattern of inheritance (represented in Fig. 7.1) in terms of the law of segregation of 'determinants' transmitted in the germ cells.

in the plants grown from those seeds or in the coats of the seeds or the seed pods of those plants. The seeds are the initial phase of the next generation and not an integral (somatic) part of the parent plant producing them. The flowers from which they are produced (by the fertilization of ovules by pollen), the stems on which they are supported, the coats and pods surrounding and containing them are all part of the parent plant.

Mendel sketched the basic explanation of the pattern his observations revealed. It is difficult today to report what he said without imputing to him notions which were stated explicitly only by his successors. However, reading between his lines may not be out of place here. He assumed that constant or pure lines can only be formed when the ovules and the fertilizing pollen are of like kind, that is, both having the character of the pure strain. He went on to assume that the ovules and the pollen of hybrids were of as many sorts as occur in their two different but pure parents. Granting the principle of dominance, it follows that a hybrid, produced when pollen and ovule were different, will manifest the dominant trait and yet produce pollen and ovules of both kinds. When self-fertilized three resultants are possible and, granted an equal production of each sort of pollen and ovules, these three will be in the proportions 1 : 2 : 1. This is presented graphically in Fig. 7.2. That is, characters which derive from the two parents and which come together in the hybrid separate again in the pollen and the ovules produced by the hybrid. This principle is spoken of as Mendel's law of segregation.

It is important to notice that Mendel used two words: trait (German, *Merkmal*) and character or disposition (German, *Character*). He employed the first for the manifest features such as round seeds, yellow cotyledons and short stems. He employed the second for the assumed determinant which was passed on from parents through the germ cells (pollen and ovules) to the offspring in which it would either produce the manifest trait or remain latent. To describe an organism in terms of its manifest traits is to specify what later geneticists have called its phenotype. To describe it in terms of its genetic potentialities, that is, the determinants it can transmit, is to specify its genotype. The pure round-seeded plants and the hybrid round-seeded plants are phenotypically the same although they are genotypically different; though they look alike, their genetic potentialities which are realized in their offspring are different.

Mendel also reported the results of experiments involving crosses between plants differing in two or three traits. Let us consider the results of the cross-fertilization of plants normally producing round yellow seeds and plants normally producing wrinkled green seeds. Mendel found that the hybrid seeds were round and yellow. Fifteen plants grown from these hybrid seeds and allowed to self-fertilize yielded 556 seeds of four sorts, sometimes all in the one pod:

(a) 315 round yellow   (c) 108 round green

(b) 101 wrinkled yellow   (d) 32 wrinkled green

When germinated the seeds in group (a) developed into (i) 38 plants producing round yellow seeds, (ii) 65 producing round yellow seeds and round green seeds, (iii) 60 producing round yellow seeds and wrinkled yellow seeds, and (iv) 138 producing round yellow seeds, round green seeds, wrinkled yellow seeds and wrinkled green seeds; the seeds in group (b) developed into (v) 28 plants producing wrinkled yellow seeds and (vi) 68 producing wrinkled yellow seeds and wrinkled green seeds; the seeds in group (c) developed into (vii) 35 plants producing round green seeds and (viii) 67 producing round green

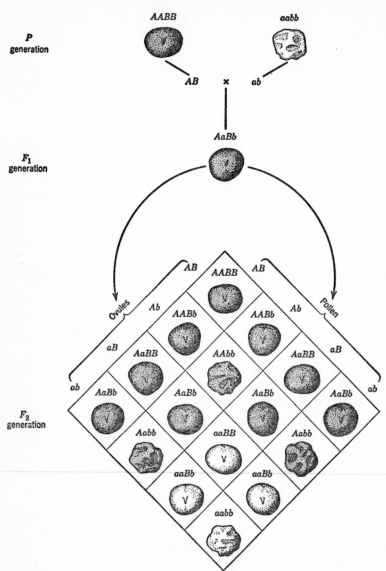

Fig. 7.3 Mendel's explanation of the pattern of inheritance of a pair of characters in terms of the law of independent assortment. (From Dobzhansky, *Evolution, Genetics and Man*, John Wiley & Sons, Inc., New York and reproduced by permission of the publishers.)

seeds and wrinkled green seeds; the seeds in group (d) developed into (ix) 30 plants producing wrinkled green seeds.

From this and other analogous experiments Mendel concluded that double hybrids produced four types of seeds (phenotypes) in the approximate proportions 9 : 3 : 3 : 1. The first type developed into plants demonstrably of four genotypes in the approximate proportions 1 : 2 : 2 : 4. The second type developed into plants of two other genotypes in the approximate proportions 1 : 2. The third type developed into plants of two further genotypes in the approximate proportions 1 : 2. The fourth type of seed developed into plants demonstrably of a ninth genotype. Mendel assumed that in these cases the characters transmitted from generation to generation segregated and combined independently of one another. This principle is spoken of as Mendel's law of independent assortment. The situation may be grasped more readily by reference to Fig. 7.3.

The supposition of a determinant transmitted through the germ cells is one of a kind demanding special consideration here. Though stating whence this determinant came and in what it resulted, Mendel did not say what it was or even where in the germ cells it was located. Theoretical notions such as this are among, or are akin to, what are called dispositional concepts or relative terms. An irascible man is not angry all the time, he is merely disposed to anger. When in the preceding case study, we spoke of metals as being ductile, malleable and good conductors of heat and electricity, we did not mean that they were being drawn out readily, were being hammered into shape readily and were conducting heat and electricity to a marked extent all the time. What was meant was that if and when they were drawn or hammered they stretched out or were shaped readily without fracturing and that if and when heat was applied at one point it was readily conducted to remote points in the metal and so on. What it is in a man that disposes him to anger is not in any way revealed by describing him as irascible. The same comment may be made in respect of ductility, malleability and conductivity. In each case, we are saying that there is a something (we know not, or we say not, what) which makes the man prone to anger or the metal ready to be drawn, to be shaped and to conduct. Speaking of determinants in the germ cells is perhaps going a little further. We know just a little more specifically where to look although we do not know any better what to look for.

Both Mendel's findings and his suggestions for an explanation of them were neglected for over thirty years. Thus the science of genetics is almost entirely a product of the twentieth century. An important part of its history in this century has been the discovery first of the location in the germ cells of Mendel's hypothetical but unspecified (uncharacterized) determinants, later called genes, and second of the probable nature of these determinants. Another part of its history has been the discovery of other parts of the variegated pattern of inheritance of manifest traits.

Sutton's and Boveri's location of
the genes on the chromosomes

In 1903 Sutton and Boveri independently of each other assumed a locus for Mendel's determinants in the chromosomes in the nuclei of cells. Cytological evidence on the chromosomes suggested this assumption.

Living organisms are either a single cell (for example, the amoeba) or a complex of many cells. The cells in multicellular organisms are typically of many different kinds, each highly specialized. For example, the many millions of cells in the human organism include skin cells, bone cells, nerve cells, muscle cells, gland cells, blood cells and so on. This complex begins with the fertilization of an egg cell by a sperm cell. The fertilized cell, or zygote, subsequently divides, its nucleus splitting to form two nuclei each becoming the control centre for one of the resulting pair of cells. This cell division continues, accompanied after a time by cell differentiation, yielding the many cells of many sorts in a multicellular organism. At certain stages in the history of each cell, a number of thread-like structures, the chromosomes, appear in the nucleus. There is a typical number of chromosomes for a given species, for example the garden pea has fourteen, the fruit-fly *Drosophila melanogasta* has eight, and man forty-six. Prior to the cell division which enables growth and differentiation the chromosomes move into a central plane. Before the cell nucleus divides, each chromosome splits longitudinally into two chromatids, one chromatid going to one new nucleus and the other to the other. Thus the pairs of chromosomes are reproduced in the daughter cells resulting from the division which is termed *mitosis* (from the Greek *mitos*, a thread) (*see* Fig. 7.4). There is another type of cell division termed *meiosis* (a Greek word meaning

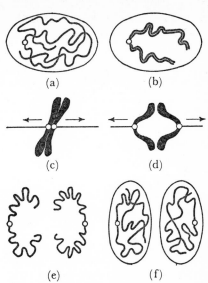

Fig. 7.4 Diagram illustrating *mitosis*. In (a) the chromosome has 'formed' after the so-called resting phase; in (b) it has split longitudinally into two chromatids held together at the centromere; in (c) the chromatids begin to move towards opposite poles in the cell, a process continuing in (d); in (e) they have moved into the emerging nuclei of the two daughter cells shown in (f).

diminution) which yields the germ cells or gametes. The chromosomes first come together in pairs—not at random but in homologous pairs. Each chromosome then splits longitudinally into a diad of chromatids held together at a central point. Instead of each chromosome contributing one chromatid to the daughter cells, as happens in *mitosis*, the diad of only one goes to the daughter cell. In the final stage of *meiosis* the diad of chromatids separate, only one going to the next daughter cell. Thus the gamete has only half the number of chromosomes present in the somatic cells. In the garden pea there are seven chromosomes in a pollen cell and seven in an ovule. When the pollen fertilizes the ovule, the nucleus of the zygote contains the full number fourteen, seven from the pollen and seven from the ovule. Sutton and Boveri saw the parallel between this derivation and pairing of chromosomes on the one hand and Mendel's notions of the segregation and combination of hypothetical genetic determinants on the other.

One implication of the location of genes in the chromosomes is that the law of independent assortment will hold only when the genes producing the separate pairs of traits are located in different chromosomes. That is, had the gene for round seeds and the gene for yellow cotyledons (and their alleles, the recessive genes for wrinkled seeds and for green cotyledons) been located in the one chromosome in the garden pea, Mendel's F_2 plants in the second of his experiments discussed here would have had either round yellow seeds or wrinkled green seeds. In that case, the manifest traits would have been linked. The number of sets of linked traits must be the number of chromosome pairs if Sutton's and Boveri's suppositions are correct. Morgan in 1916 provided an impressive confirmation by showing that there are four sets of linked manifest traits in *Drosophila melanogasta* which has four pairs of chromosomes. The number of independent sets of manifest traits in the garden pea with its seven pairs of chromosomes must similarly be seven. Thus unless Mendel had remarkable insight into the genetic behaviour of his chosen organism, he had remarkable luck when he selected the seven pairs of traits he did. In terms of the Sutton-Boveri hypothesis, the genes for each pair were located in a separate pair of homologous chromosomes.

Quite early it became clear that the linkage of manifest traits was not complete. In 1906, Bateson and Punnett reported a case of partial linkage in sweet peas. The traits concerned were flower colour, blue being dominant to red, and pollen grain shape, long being dominant to round. Pure, that is homozygous, blue-long plants were crossed with homozygous red-round plants, yielding an F_1 generation of blue-long plants. Let us denote the dominant gene for blue flowers *B* and its recessive allele for red *b*, and the dominant gene for long pollen grains *L* and its recessive allele for short *l*. If the law of independent assort-ment held here, these hybrids would produce four kinds of gamete, *BL*, *Bl*, *bL*, and *bl*. If the genes *B* (or *b*) and *L* (or *l*) were located in the one chromosome, then one would expect two kinds of gamete, *BL* and *bl*. Back-crossing these hybrids (heterozygous plants) with homozygous recessive plants of the *bbll* genotype, should, in accordance with these two possibilities, yield *either* (i) in terms of independent assortment, four phenotypes in equal proportions—there will be four gametes *BL*, *Bl*, *bL* and *bl* in equal numbers being combined with a *bl* gamete, thus there will be 25 per cent *BbLl*, 25 per cent *Bbll*, 25 per cent *bbLl* and 25 per cent *bbll*, each phenotypically different, namely blue-long, blue-round, red-long and red-round, *or* (ii) in terms of complete linkage, two phenotypes in equal proportions—there will be two gametes *BL* and *bl* in equal numbers being combined with a *bl* gamete, thus there will be 50 per cent *BbLl* and 50 per cent *bbll*, each phenotypically different, namely blue-long and red-round.

Bateson and Punnett found four phenotypes as in (i) but they were

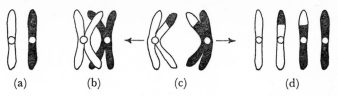

(a)   (b)   (c)   (d)

Fig. 7.5   Diagram illustrating cross-over in *meiosis*. In (a) a pair of homologous chromosomes have lined up together; in (b) they have each split longitudinally to form two chromatids, a crossing-over of two of these chromatids is shown; in (c) when the diads of chromatids move apart, the material beyond the cross-over is shown as being exchanged; and in (d) the diads of chromatids have separated—in spermatogenesis each goes into the nucleus of a gamete and in ovagenesis one of the four goes into the nucleus of a gamete, the other three being rejected.

not in equal proportions. The two phenotypes expected in (ii) appeared most frequently, but the other two phenotypes expected in (i) also appeared occasionally. The proportions were blue-long 43.7 per cent, red-round 43.7 per cent, blue-round 6.3 per cent and red-long 6.3 per cent. When homozygous blue-round (*BBll*) plants were crossed with homozygous red-long (*bbLL*) plants, and the hybrids back crossed with homozygous red-round (*bbll*) plants, the four phenotypes appeared again in the F_2 generation in unequal proportions though the previously infrequent ones were now the frequent ones. Thus the blue-long and red-round each occurred with a 6.3 per cent frequency, whereas blue-round and red-long each occurred 43.7 per cent.

A cytological explanation of such incomplete linkage was provided by Morgan on the basis of some evidence obtained by Janssens. In the early stages of *meiosis*, the cell division which ultimately produces a gamete, the chromosomes line up in their homologous pairs and split longitudinally to give four chromatids in place of the original two chromosomes. Janssens suggested that sometimes two chromatids of the pair of homologous chromosomes bend across each other, break at the cross-over point and rejoin so that segments of the two chromatids are exchanged. This process is illustrated diagrammatically in Fig. 7.5. Were the genes *B* and *L* (or their alleles *b* and *l*) some distance apart in the chromosomes, such cross-overs would occasionally break the linkage of the two manifest characters, blue colour of the flower and long pollen grains.

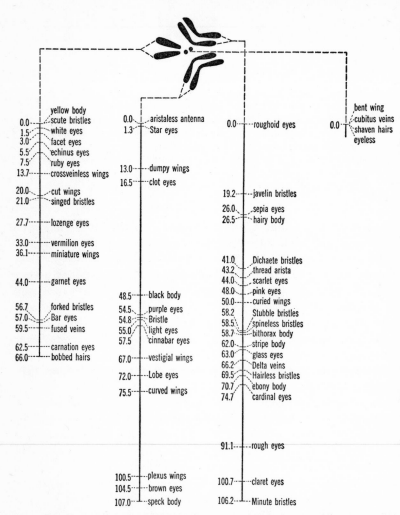

Fig. 7.6 Gene maps of the four chromosome pairs of *Drosophila melanogasta*. (From Dobzhansky, *Evolution, Genetics and Man*, John Wiley & Sons, Inc., New York and reproduced by permission of the publishers.)

Such exchange of segments between homologous chromatids would result in the partial breaking of a linkage more frequently where the genes concerned have widely spaced locations than where they are close together. When widely spaced any one of a fairly large number of points of cross-over would serve to separate them; when close together, only a small number of possible points of cross-over would separate them. By analysing the frequencies with which various linkages were broken, Morgan was able to produce gene maps for each of the four pairs of chromosomes in *Drosophila melanogasta* (*see* Fig. 7.6).

Other patterns of inheritance

Further experiments and analyses of data derived from other organisms revealed how complex the total pattern of heredity is. Just a few extensions of what has been reported above will be mentioned in illustration. Any reader who is interested in a fuller account should turn to some systematic exposition of genetics.

The cases so far reported have consisted of a dominant trait paired with a recessive. Two variations from this pattern will be reported. First, there are cases where the manifest trait in the hybrid is a blend of the manifest trait of the parents. For example, with shorthorn cattle, a red-white cross produces hybrid offspring with a roan coat, that is their coats consist of interspersed red and white hairs. Second, there are cases which can be explained in terms not of a pair of alleles but instead of a larger set of alleles. For example, there are four human blood groups (phenotypes) A, B, AB, and O, defined originally in terms of compatibility in respect of inter-individual blood transfusion. These phenotypes are produced by three alleles A and B, which have no dominance relation to each other, and O which is recessive to the other two. Phenotype A is genotypically AA or AO; phenotype B is genotypically BB or BO; phenotype AB is genotypically AB; and phenotype O is genotypically OO.

In all the above cases one gene and its allele (or alleles) are the determinants of the manifest trait. There are also cases of several genes, sometimes in different chromosomes being involved. Two cases will be cited. The first is the genetic determination of coat colour in horses and the second, a more speculative case, of skin colour in human beings. Let us begin with bay, black, and chestnut horses (*see* Fig. 7.7). Two pairs of genes are involved, B which is dominant to b and I which is

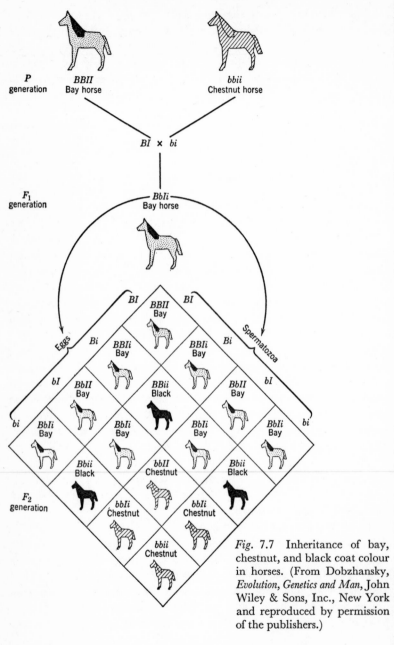

Fig. 7.7 Inheritance of bay, chestnut, and black coat colour in horses. (From Dobzhansky, *Evolution, Genetics and Man*, John Wiley & Sons, Inc., New York and reproduced by permission of the publishers.)

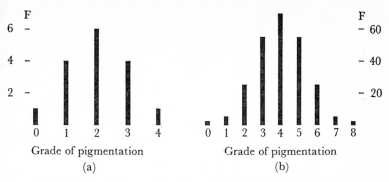

Fig. 7.8 Distribution of grades of skin pigmentation on the assumption of (a) two pairs of pigmenting alleles and (b) four pairs of pigmenting alleles.

dominant to *i*. Chestnuts are homozygous only in respect of the recessive *b*; that is, genotypically they may be *bbii*, *bbIi* or *bbII*. Blacks are homozygous in respect of the recessive *i* and have at least one dominant *B* gene; that is, genotypically they may be *BBii* or *Bbii*. Bays have at least one dominant *B* and at least one dominant *I*; that is, genotypically they may be *BBII*, *BbII*, *BBIi* or *BbIi*. All three are homozygous in respect of a recessive *d*. The dominant *D* converts what would otherwise be a bay horse into a dun coloured horse, and what would otherwise be a black into a bluish 'mouse' coloured horse. A further dominant *G* produces scattered white hairs which increase with age; such horses become greys in old age. There is another gene which in the homozygous case gives white (a white skin as well as uniform white hairs) and in the heterozygous case the palomino coat. Another gene, when dominant, produces pied coats.

Human skin colour may be determined by several pairs of alleles not in a dominance relation and located in different chromosomes. It is suggested that a gene *P* contributes to pigmentation of the skin and *p* not. Thus if there were two pairs of pigmenting alleles, $P_1P_1P_2P_2$ would produce a very dark skin and $p_1p_1p_2p_2$ a very light skin, with several possible grades between, as is shown in (a) of Fig. 7.8. A finer gradation of phenotypes is possible with four pairs as is shown in (b) of that figure. Such arrangements yield binomial distributions of pigmentation which, when the number of independent determinants becomes large, approximate the mathematical normal distribution.

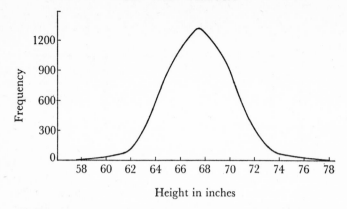

Fig. 7.9 Distribution of standing height of 8,585 adult British males. (Reproduced from G. U. Yule and M. G. Kendall, *An Introduction to the Theory of Statistics* (14th edition, 1965), Charles Griffin & Co. Ltd, by permission of author and publisher.)

Many human traits affected in part, if not entirely, by heredity are distributed in an approximately normal way. One example is stature, the distribution of which is depicted in Fig. 7.9.

Another variant in the pattern of inheritance is attributed to mutation, a change in the gene. Some recessives, for example that producing haemophilia, tend to be lethal. As many of the individuals manifesting the trait would die before reproducing themselves, one would expect such genes to become rare if not to disappear in due course from a population. That they do not do so suggests that they crop up in some 'spontaneous' manner from time to time. Morgan's extensive experiments with *Drosophila* strongly suggest such mutations. Later it was shown that radiation will produce changes in genetic lines which seem explicable only in terms of mutations. There is more than a hint that a set of alleles derive by mutation from a single original gene.

On the basis of the findings presented above the gene came to be thought of as a discrete entity, a particle in the chromosome. After gene mapping had been introduced it was sometimes said that the genes were like beads on a string, each string of beads being a chromosome. When suitable microscopic techniques showed that the chromosomes were banded, it was even suggested that these bands might be the genes or, at any rate, sets of genes. However, some of the more critical geneticists considered such a particulate conception to be an over-simplification of the complex theory for which the explanation of the facts called.

The DNA molecule

Completely new light was thrown on the possible nature of genes by Watson and Crick in 1953 when they suggested what may be the structure of the deoxyribonucleic acid molecule. Chromosomes are composed chiefly of three chemical substances: proteins, deoxyribonucleic acid (DNA), and ribonucleic acid (RNA). All three are polymers which are large molecules formed by the repetitive arrangement of less complex molecules. Proteins consist of anything from 100 to 10,000 units of some twenty amino acids. The amino acids consist of a carboxyl (COOH) and one or more amino groups (NH_2) attached to a carbon atom together with other atoms or molecules. A typical formula is

$$NH_2$$
$$|$$
$$R.CH.COOH$$

where R stands for an atom or set of atoms, for instance in glycine the additional atom is H, and in alanine the additional atoms are CH_3. This is illustrated in the following five amino acids.

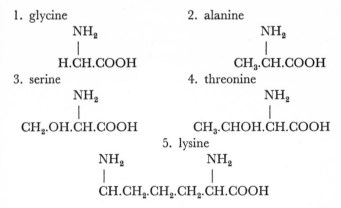

1. glycine

$$NH_2$$
$$|$$
$$H.CH.COOH$$

2. alanine

$$NH_2$$
$$|$$
$$CH_3.CH.COOH$$

3. serine

$$NH_2$$
$$|$$
$$CH_2.OH.CH.COOH$$

4. threonine

$$NH_2$$
$$|$$
$$CH_3.CHOH.CH.COOH$$

5. lysine

$$NH_2 \qquad NH_2$$
$$| \qquad |$$
$$CH.CH_2.CH_2.CH_2.CH.COOH$$

When hundreds and even thousands of such 'building blocks' are put together to make a protein molecule the latter is a very complex structure. DNA and RNA are even more complex. They consist of

145

repetitive linkages of four nucleotides. Each nucleotide consists of a phosphoric acid, a sugar and a nitrogenous base. In DNA, the sugar is deoxyribose, and the bases are two purines (adenine and guanine) and two pyrimidines (cytosine and thymine). Watson and Crick have suggested that each DNA molecule consists of two intertwined helical strands of phosphate-sugar chains linked by pairs of bases much in the manner that the strings of a spiral staircase are linked by its treads. Each tread in this staircase, they suggested, consists of either adenine and thymine or guanine and cytosine. As each base in the tread could be on the left or on the right, four different treads may occur in the staircase and, of course, in a great variety of sequences. This hypothetical structure of the DNA molecule is presented in Figs 7.10, 7.11 and 7.12.

Two features of this hypothetical structure of DNA need to be stressed in the present context. First, the sequence of the treads in the staircase permits the coding of information so that those who climb may read. Just as the two units of Morse code, dot and dash, enable a large variety of messages to be coded provided one can arrange the units in various orders in sequences of considerable length, so the four letter alphabet of the Watson-Crick model provides for dis-criminable messages of many thousand letters in length. There are not enough letters in this alphabet for genes to be one letter words. One could not say much if one had only the four words 'ay', 'bee', 'cee' and 'dee' which had to be used in isolation. However if one can combine them as letters to make words (or sentences) varying from one letter to 15,000 letters (the number of nucleotides in some nucleic acid chains), then numerous quite complex pieces of information can be

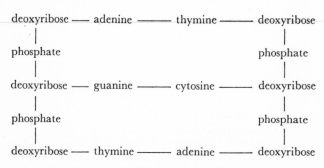

Fig. 7.10 The Watson-Crick model of the DNA molecule.

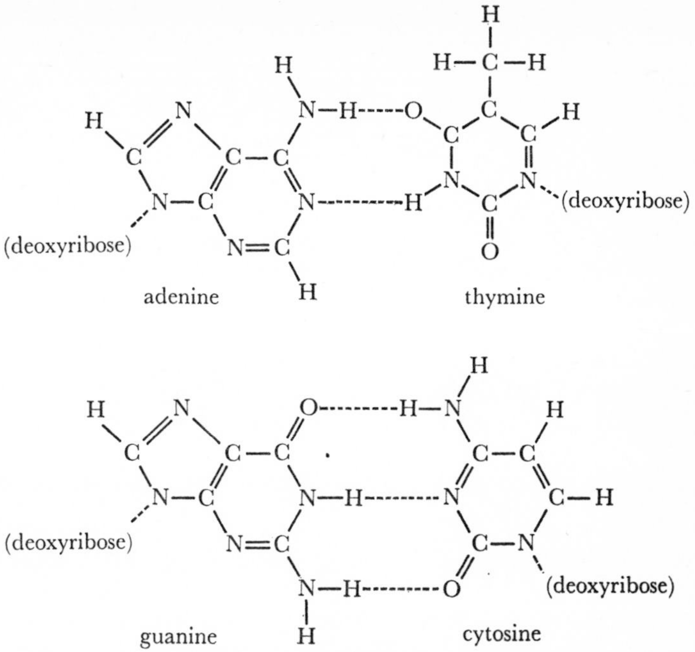

Fig. 7.11 Structure of the link (tread) of two bases between the sugar (deoxyribose) molecules in DNA.

transmitted. The 'message' would thus not be at one point ('tread') in the staircase and it may be gained sometimes only by ascending a section of a staircase and sometimes only by ascending several staircases. Second, should the DNA molecule as envisaged by Watson and Crick split down the middle, there is enough information on one half to enable the reconstruction of the other. Each adenine needs a thymine for the completion of the tread, each thymine an adenine, each cytosine a guanine and each guanine a cytosine. The relevance of this for mitotic cell division is plain. It is significant that during what used to be called the 'resting' phase after the new cells are formed, the quantity of DNA in the nuclei doubles.

Though much of this story needs to be confirmed and though little more than the opening sections of it have been told, it would seem that we now know not merely where genes are located but also what they are. Just how what is coded in the DNA molecule is realized in

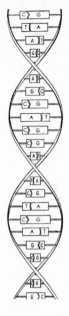

Fig. 7.12   Diagram showing the helical form of the DNA molecule as depicted by Watson and Crick. (From David M. Bonner, *Heredity*, © 1961. Reprinted by permission of Prentice-Hall, Inc., Englewood Cliffs, New Jersey.)

the manifest traits of the organism is not yet fully, perhaps not even well, understood. There are hints that RNA acts as a messenger carrying the 'instructions' coded in the DNA molecule to the sites of differentiation elsewhere in the cell, and that the 'instructions' are for the production of enzymes which act as the templates for the building of the various protein molecules. How any sequence of treads in the DNA molecule is to be translated in terms of manifest characters is a matter for the future to reveal. That is, there is no way open yet for the deduction of manifest traits from the sequences of bases in the DNA molecule, as there is a way of deducing many of the properties of elements from a given pattern of electrons in the shells of the atom.

A review

It has already been pointed out that Mendel's first intimations of the gene provided a purely dispositional concept. His theorizing about it was also largely *ad hoc*. The facts of inheritance of certain differentiating traits in the garden pea were so and so, and the laws of segregation and of independent assortment and the notions of dominant and recessive determinants were stated in ways that would directly fit these facts. Knowing the laws and which traits were paired in a dominant-recessive relationship, one could infer the facts. The premises of the inference had been chosen to enable just these conclusions. They did not, however, enable conclusions about other observable situations. Mendel's theory in this respect was like the theories of Eudoxos and of Ptolemy.

In giving the genes a location in the chromosomes, Sutton and Boveri made the theory testable in a partial way. By recognizing the Earth's orbit in the deferents of Mercury and Venus and in the epicycles of Mars, Jupiter, and Saturn, Copernicus provided a means of testing the *ad hoc* elements in his heliofocal theory, elements which, like those in Ptolemy's geostatic theory, had to be introduced in order to infer celestial positions from the theory. By recognizing that the Earth was in motion he provided through the implied stellar parallaxes a test, admittedly one of great technical difficulty, of that central feature of the theory itself. The check on *ad hoc* theorizing provided by Sutton and Boveri was not as rigorous as those provided by Copernicus. This is shown by the adjustments made in the extended theory when linkages in many instances proved to be partial rather than complete. Had linkages always been complete, the extended theory, that is, Mendel's plus the Sutton-Boveri hypothesis, would have been confirmed. Had they been quite absent then the extended theory would have been falsified. In the latter event it is almost certain that only the Sutton-Boveri extension which had enabled the test would have been abandoned. That linkages were partial but varying in degree enabled theorists to speak of the spacing of gene loci and to engage in gene mapping in a consistent and plausible way. These adjustments, however, had some of the flavour of the adjustments to the Ptolemaic theory. Nevertheless, the whole thing was no longer completely *ad hoc*. For instance, it is hard to see what tinkering could have been effective had there been sets of linkages, partial or complete, outnumbering the chromosomes.

Had Janssens' evidence for exchanges between the chromatids at cross-over points been less speculative, then Morgan's attribution of partial break-down of linkages to them would also have been less like *ad hoc* tinkering. Of course, the fact that Janssens or any one else with his observational facilities could only see the chromatids from homologous chromosomes cross but could not actually see them break at the cross-over point and re-form with exchanged segments did not require that no one ever would. This lack of observation was not the result of an inherent unobservability. It was something analogous to the expected but elusive stellar parallaxes which afforded an important test of the heliofocal theory. Thus Morgan's assumption of crossing over (more strictly, exchange of chromosomal material) was not in the same class as Ptolemy's equants which could only be checked through the phenomena they were introduced to explain.

Like Bohr's theory of sub-atomic structure, which was not developed explicitly to explain the periodicity in the properties of the elements, the Watson-Crick hypothesis of the structure of the DNA molecule is based on chemical evidence and not on genetic considerations. Nevertheless it gives promise of explaining genetic facts. However, whereas Bohr's theory has already a vast achievement and a shrinking area of mere promise, the Watson-Crick hypothesis is genetically speaking largely a matter of promise. It makes sense of the broad pattern of brute facts, but it does nothing yet to make sense of any of the details. No one yet knows how to read the genetic code assuming it is programmed on the treads of the DNA molecule, so that no phenotypic predictions from the theory are possible. For that matter no one can yet say precisely how the 'instructions' coded in the molecule are carried out in cell differentiation. This, of course, is probably a mark of immaturity in this line of theorizing rather than of some inherent defect in it. We may come to reject the view that there is a relation between the structure of the DNA molecule and the genetic determinants which Mendel glimpsed. If we do, it will be either because elaboration of the theory and the expansion of the facts show it and them to be incompatible or because another theory will offer a more comprehensive or a more detailed explanation of the relevant facts.

8 COMMENTS

THEORETICAL ENTITIES AND
REDUCTIVE EXPLANATIONS

The view that theories are not to be taken as literally asserting what obtains in nature is frequently put in relation to such theoretical entities as protons, electrons and genes. It is said that unlike tables and chairs, planets and crystals of common salt, and the pointing of needles to marks on the dials of measuring instruments, theoretical entities are to be taken not as real things but instead as fictions or conventions useful in thinking about the ways in which nature works. It is said that they are like a scaffold, useful for the workman on a building but not part of that building. Before we can examine these contentions we shall need to make some distinctions between entities and properties on the one hand and relations on the other.

Terms and relations

Both facts and theories contain references to things or entities, to the properties or attributes of things, to relations between things or between properties, and to situations. A proper disentangling and classification of these several referents would involve a long and intricate incursion into logic and probably into metaphysics as well. It would also throw up a great deal which would prove to be highly controversial without necessarily clarifying any further the issues under discussion. Some discussion, no matter how superficial, nevertheless, is necessary for the understanding of some points that have yet to be made about theories, especially those involving contentions about assumed entities and those involving contentions of the type called scientific laws or laws of nature.

What we ordinarily regard as a thing is or has a distinctive set of properties and it has some historical continuity. A freshly-minted penny is a thing. It is a disc of such and such width and thickness, it has a given weight, it is made of such and such a type of bronze, it has a patterned texture on both its sides, it can be used to pay for goods and services and so on. We may be inclined to think of these properties as clinging to some substance but it is difficult to say what that substance is without citing these or other properties. We need not pause here, however, to settle whether the penny *has* these properties or *is* these properties. When the sheen wears off the freshly-minted penny and its lustrous orange turns to dull brown, we do not contend that it has turned into another penny. The old penny is historically continuous with the new; it is merely that some of the properties have changed with the passage of time. Though the penny has continuity with a piece of unminted bronze, we do not regard the latter as a penny. The piece of metal was turned into a penny by changing a particular set of properties. It ceases to be a penny on being melted down again. The minimal set of properties constituting a thing is distinctive of it. An entity involves identity. Some properties may change without affecting that identity, but there is a minimal set which provides it.

Properties or attributes are of many varieties. There are for instance 'sensory' qualities such as red, loud, sweet, and pungent, spatial qualities such as straight, square, and cylindrical, and quantities such as frequencies and amounts. These may be combined in various ways such as 'dual red squares' or 'slightly sweet white crystals'. The examples and their grouping here are not meant to be exhaustive; they are given merely to illustrate the variety of properties.

Relations are the way things, or properties, stand to one another. Relations obtain between things, or between properties. In saying that this square is larger than that, we are pointing to a quantitative relation between the areas of the two squares. In saying that the disc of Jupiter is oblate, we are pointing to the relation of predication (affirming or denying a property of a subject) between being oblate and Jupiter's disc. In saying that genes determine such manifest characters of plants as seed shape, we are relating genes (theoretical entities in the chromosomes) causally with the given characters.

It should suffice here if we group things and properties as terms and distinguish them from relations, which hold between terms. A situation may then be described as a set of terms-in-relation taken as a whole. Facts and theories assert situations, observed and assumed. Each of the following statements asserts a situation, a set of terms in relation— 'the sidereal period of Mercury is 88 days'; 'an electric current is a flow of electrons through the network of atoms making up a conductor'; 'the genes for flower colour and for shape of pollen grain in sweet peas lie in the one chromosome'.

Where the situation in such cases is observed as a whole, that is, both in all its parts and in its entirety as asserted, we speak of the statement as a fact. Where there are assumed terms or assumed relations, we speak of the statement as a theory. As examples of assumed terms may be cited the hypothetical Vulcan assumed by Leverrier, the sub-atomic particles, protons, neutrons and electrons, assumed by Bohr following the lead of Rutherford, and genes assumed explicitly by Mendel's twentieth-century successors. In Newton's notion of gravitation the assumption is of an attractive force between bodies of matter and of the variation of this force in direct proportion to the produce of the masses of the bodies and in inverse proportion to the square of their distance apart. The terms, masses and distances, may in many situations be observed; one may observe also signs of the relation of attraction (for example in hefting a stone) but it is more precise to say that the attraction is assumed. Though one may test the law

$$F = G\,m_1\,m_2\,/\,r^2$$

by making observations, the complex of relations between the observed terms, m_1, m_2, and r, is assumed.

Often when a hypothetical term is introduced, it cannot be positively characterized. The theorist using it says how the theoretical entity works or how it is related to other things though he cannot say what it is, what set of qualities it is or has. To say that Jupiter's disc is oblate and that the seeds of certain varieties of garden peas are round is to characterize these subjects positively. When Adams and Leverrier introduced the hypothetical Neptune they characterized it and its orbit positively; for example, each said that it had a specified mass, that its sidereal period was so many terrestrial years and so on. They also stated a number of relations, for example, that it stood at such and such times in such and such a direction relative to Uranus from the Sun, and at such and such distances from Uranus. When the notions 'positively charged' and 'negatively charged' were first introduced to describe the electrical state of particles, they were not characterized in terms of qualities. They were references to the disposition of certain particles to attract or to repel each other. That is, they were relative or dispositional terms. When genes were introduced they were thought of as entities of some sort but what the sort was could not be said. They could be thought of only dispositionally, that is, as determinants, derived from the parents, of given manifest characters in the offspring.

Sutton's and Boveri's location of genes in the chromosomes was a first step in the qualitative characterization of genes.

Theoretical terms which remain purely dispositional are of very limited value, indeed one may say, subject to a later qualification, that any value they have resides in the possibility that they may in future be qualitatively characterized and then shown to have genuine explanatory uses. On the other hand there are dangers in the too ready endowment of hypothetical terms with properties when there is not some observational support for or check on the attribution. Late nineteenth-century physics ran into embarrassments through too great a readiness to endow the assumed medium aether with properties, the endowment being intellectually satisfying rather than empirically testable. In this century an extreme positivist prophylactic against such prodigality has been proposed, namely that a term should be deemed to be synonymous with, to refer to no more than, the operations employed in assessing or in bringing about its referent (property, thing or situation). Thus an operational definition or specification of length should, it is said, not go beyond what one can observe by applying a foot-rule in some standard manner to a physical object, and an operational specification of an electric current should not go beyond pointing to what one observes when one works a generator (specified in certain ways) or when one applies and reads instruments such as voltmeters and ammeters which may be qualitatively characterized. There is a salutary warning in this operationist rule but it is rather too restrictive a guide to all that science involves. It urges the scientist, in effect, not to take the risks incurred in moving far from the facts. However, it may be properly asked whether science can be undertaken without taking the risks of skating on the possibly thin ice of supposition. The important thing is to know when one is on the more solid ground of observation and when one is on the ice.

Some terms important in science and some relations between terms are the direct product of observation and others are largely the product of assumption. The dividing line between the two, as has been repeatedly said about fact and theory, is fine and wavering. Night by night observations from the Earth of the position of Mars among the stars seem to leave no doubt that the real path of Mars forms a closed figure, whatever the precise form of the figure be and whether it be focused on the Earth or on the Sun. However, one is prompted to admit that there may well be some assumption here when one compares Mars in this respect with Mercury. Though early observations suggested otherwise, later more refined observations indicated that Mercury's orbit is not closed but instead that the planet, because of the progression of its perihelion, never comes back to the point from which it begins any given circuit. Were we able to make sufficiently precise measurements of Mars' positions, the same might (and probably would) be found of it. Nevertheless, even granted that Mars' orbit is a closed figure, the form of the figure remains to be established. Kepler's proposal that the figure is an ellipse with the Sun in one focus seems the best simple approximation. This is, of course, primarily an assumption even though it is meant to tie together in a simple yet adequate way a large number of observed data.

When atoms were bombarded by alpha radiation in an apparatus spoken of as a cloud chamber and occasional forked vapour trails were observed, it was tempting to say that one or other of the sub-atomic particles assumed by Rutherford and by Bohr had been observed as Adams' and Leverrier's hypothesized Neptune had been observed by Galle. Though the particles themselves may not have been observed, signs of them had been. One may reasonably, if not with complete justification, say that one can see the vapour trail of a high-flying aeroplane even though the aeroplane itself is out of sight. The grounds making this a reasonable statement are first, that one has on other occasions seen high-flying aeroplanes making trails just like this one and second, that one has never seen anything else making a trail like it. But it is one thing in everyday life to say that one has seen goat tracks at the bottom of one's garden, even though one has never caught the marauders amongst one's cabbages, and quite another to identify other depressions there as leprechaun tracks (granted that one is not among the privileged few who have witnessed a covey of the little folk circling on soft ground). Unless one has seen on some occasion both the flying particle and the trail it makes, one cannot on other occasions say that one has observed the flying particle through its signs or claim that what one has seen is an indubitable sign of it. One may, of course, say that one has obtained empirical confirmation for the assumption of sub-atomic particles because in the hypothesis the particle has been characterized by properties such that were it to move with high speed through a cloud of water vapour it would produce an observable vapour trail. But confirming a theory and observing the situation it assumes, or some component of that situation, are different matters.

The observation of stellar parallaxes has confirmed the heliofocal theory but it has not been an observation of the Earth's revolution around the Sun.

It is at this point that one may emphasize the difficulty with hypothetical terms defined only in an operational or a dispositional way. The only states of affairs one may infer from the assumed presence of such an entity are the sort which the entity is assumed to be disposed to bring about. It may be a convenient manner of speaking to say that there is a pair of somethings in the zygote, derived one from each parent, which stand in a dominant-recessive relation to each other and which are disposed to bring about long stems and short stems respectively in the mature plant. It may even contain the promise of an explanation but that promise can only be fulfilled when someone is able to characterize those entities through the attribution of properties to them which enable either their identification by observation or their indirect confirmation through the observation of other sorts of situations implied by them.

Recognizing this limitation of theoretical terms specified only in a dispositional or relative way, some philosophers of science have stressed the importance of relating the terms to observable antecedent as well as to observable consequent events. To a degree this is what Mendel did with the latent characters (or genes). He related them back to the manifest traits of the parents and forward to the manifest traits of the offspring. However if one ties such terms securely at both ends to observable events, one might just as well speak of relations between the two sets of observed events without interposing any hypothetical events between them. The hypothetical terms may be convenient in speaking of and thinking about the situation but they do not add much, if anything, of substance to the two sets of observed terms and the relation between them. In that event such terms may justifiably be regarded as convenient fictions.

Few would object to the contention that some terms in science are observed whereas others are assumed. Some philosophers have argued, however, that relations between terms are never observed but that instead they are supplied in some way by the beholder. This argument is based sometimes on the wider view that reality occurs in discrete, unrelated atoms and sometimes on the psycho-physiological view that the sense receptors which mediate observation are punctate in structure and so are sensitive only to the 'sensory' qualities of objects and not to any relations between these qualities. The latter view will not resist much scrutiny as it restricts our direct knowledge of the world around us to 'sensory' qualities and to quantitative variations in them. Spatial qualities would escape the senses including the tactile senses. The former view is a half-way house to subjective idealism. Though it recognizes, as realism does, that we have direct knowledge of qualities (or terms), it also admits, as idealism does, that seeming knowledge of relations between such terms is supplied by the beholder. It is beyond the scope of this discussion, however, to go into such epistemological issues. Suffice it to say that when we see a large circle surrounding a smaller one we also see that the former is larger than the latter; that when we see two distinguishable patches of red side by side we see one as lighter than or as more orange than or as different in shape from or as to the left of the other. Just that is the minimum claim made here when it is said that on some occasions we observe the relations between terms. Or to put it the other way, it is denied that we observe only the terms and always supply by some process of supposition whatever relation is 'seen' to hold between them. It may be said as a piece of psychological information that terms observed in the proximity of time and space are always observed as being in some

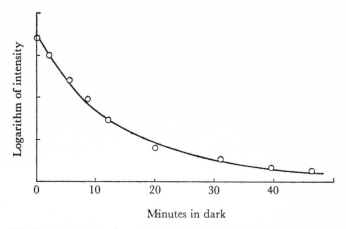

Fig. 8.1 Curve fitted to data obtained by Piper on sensitivity of the eye after varying periods in the dark. (From S. Hecht, in Murchison, Carl Allanmore, Editor, *A Handbook of General Experimental Psychology*, [Clark University Press, 1934], New York: Russell & Russell, 1969.)

relation to one another. However, some relations which are important for science have to be supplied by hypothesis.

In order to explore whether the origin of one class of relations is observational or hypothetical, we shall consider the increased sensitivity of the eye to dim light after the eye has been shielded from brighter light. It is well known that when one passes from a well-lit to a poorly-lit room one can see very little if anything of the new surroundings. After a short time, however, more and more becomes visible. It is not this observation of some positive functional relation between length of time in the dark and degree of sensitivity that is in question. What may be doubted is whether the precise form of the relation is observed or assumed. A standard experiment may be carried out as follows: a subject, after being exposed for a given time to illumination of a given level, is kept in darkness for varying intervals of time and then tested to establish the minimum intensity of light to which he is sensitive. A typical result, obtained in 1903 by Piper, is shown in Fig. 8.1. It will be noted that the plotted points, each representing the result of a test, or the average of several tests, after a given interval in the dark, lie approximately on a curved graph of a kind termed an exponential curve for which the basic equation is

$$y = a^x$$

where y is the minimum light intensity to which the eye is sensitive, x the length of time in the dark and a an empirical constant. In the present case as the graph is a descending curve the index, x, must be given a negative value.

It is easy to think that the functional relation reported by Piper was directly established by observation. However, it should be noticed, first, that the plotted points do not all lie on the curve which it is said depicts the relation. The curve at best approximates to the plotted points. The departure of the points from the curve in several instances may be attributed to random errors of observation. This suggestion may be confirmed (or falsified) by making many more tests at each period of elapsed time and averaging. If the departures are due to random errors, as the number of observations contributing to the average increase the degree of the departures would decrease. This mode of explaining away the actual departures makes clearer that the curve is at least part hypothesis and not pure observation. Second, other lines could be fitted to the data. Some simpler ones, for example, the straight line, $y = a + bx$, would not fit as well as $y = a^x$, whereas

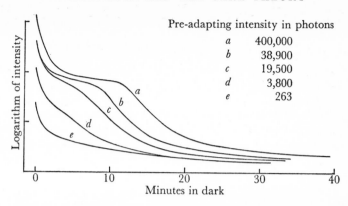

Fig. 8.2 Curves fitted to data obtained by Hecht, Haig and Chase on sensitivity of the eye after varying periods in the dark following varying levels of pre-adaptation illumination. (From *Journal of General Physiology*, Vol. 20 (1937), p. 837.)

some others which fitted better would be mathematically forbidding. Some of the complex alternative graphs would be likely to be falsified by further observations, but an embarrassingly large range of alternative possibilities to the curve drawn would be left if we did not resort to some criterion as giving preference to the simplest graph having the best fit, or the large number of not markedly different simpler graphs giving very good fits to the data. This criterion is, of course, arbitrary even if convenient.

It is worth following up Piper's early observations with some later ones made by Hecht, Haig and Chase in 1937 and displayed in Fig. 8.2. Here the intensity of the illumination preceding the period in the darkness was varied. The data following very bright pre-adaptation illuminations strongly suggest dual phases in dark adaptation, whereas the data following very low pre-adaptation illuminations do not suggest a dual process. Some of the departures from the monotonic curve fitted to Piper's data may well be faint signs of the duality more apparent at higher pre-adaptation illumination. We have, as it happens, a rationale for the dual phases which show up in the data from Hecht, Haig and Chase's experiment. There are two types of receptor in the retina of the eye which seem to have markedly different adaptation characteristics. Those which can adapt to the greatest extent seem to be slow starters when they have been exposed to high levels of illumination.

The point, however, that is to be made is that contentions about rates of dark-adaptation are far from having observations as sole bases: rather they are contentions with a substantial hypothetical component.

Convenient fictions and uncomfortable realities

We are now in a better position to discuss the issue whether theories, in so far as they go beyond the facts of observation, are to be regarded as conjectures as to what really obtains in nature or as convenient modes of thinking about what has been observed in nature. Bellarmino was willing to let Galileo say that the Earth moved around the Sun provided he added that this was just a manner of speaking which made for easier analysis and more effective inference. Bellarmino's was an early version of the instrumentalist view which holds theories to be neither true nor false but instead more or less effective instruments in the hands of the student of nature. Bellarmino suggested in effect that they are not contentions about the fabric of the body of natural phenomena but are scalpels and forceps in the hands of nature's anatomist.

In some cases, at least, there is no doubt what answer the scientist will give to the question: 'Is what you are supposing alleged to be something which really exists or is it only a manner of speaking and reasoning aimed at arriving at sounder conclusions about observable situations?' Not only did Adams and Leverrier, both theoreticians, want an astronomer who was experienced in observing to look in the sky for their hypothetical Neptune but Galle took Leverrier seriously enough to look with success. Historians of astronomy usually take to task the British astronomer Challis for his dilatory response to Adams' request to look for the hypothesized planet. Challis showed as little diligence as he did, if he were not too preoccupied with other tasks, not because he regarded Adams' supposition as a fiction convenient for shoring up the Newtonian theory but because he doubted that there was any extra-Uranian planet to be seen. Again, if Leverrier's assumption of Vulcan were only a fiction meant to prop another sagging beam in Newton's theoretical structure, it would not matter that no one, perhaps not excluding Lescarbault, has ever observed anything that could be justifiably identified as an intra-Mercurial planet. Again, a consideration of Sutton's and Boveri's hypothesis that the genes are located in the chromosomes and Morgan's hypothesis of chromosomal cross-over as an explanation of partial linkages suggests that these scientists believed that genes existed and were not merely useful conventions with which to draw inferences about patterns of inheritance. While genes were purely dispositional, an instrumentalist view of them could be plausibly maintained. But once it was suggested where they were to be found and in what order they were arranged, it could no longer be suggested that their proponents need not regard them as real. A similar argument can be developed in relation to protons, neutrons and electrons.

It is important to notice a difference between Neptune and Vulcan on the one hand and genes and the sub-atomic particles on the other. The former were merely additional instances, though assumed ones, of a kind of thing already observed elsewhere. The latter, however, are things of a sort never observed elsewhere. It is not that some genes had been observed and some additional ones assumed. We can not yet be certain that genes have ever been observed. However should it become more certain that several treads in the DNA staircase play the role that genes were invented for then it may be possible to say that genes have been observed, granted that it can be properly said that we observe rather than suppose the structure of complex molecules. The case for the observability of the sub-atomic particles is probably harder to sustain. Perhaps the best we can say is that signs of them are observable but that, as was argued above, we know these to be their signs only on theoretical grounds. That might seem to contain an admission that it would not matter whether the fundamental particles really exist in nature or whether nature worked *as if* they existed. Some comments on this suggestion will be made below when working hypotheses are being discussed. For the moment, however, it may be said that scientists act as though hypothesized entities (terms) were real. They do so in the way they test such hypotheses. They are not merely concerned to show that they serve the purposes for which they were invented but also explore avenues that could lead to their falsification. It might be shown that Eudoxos' entities (the homocentric spheres) will work, provided that hypothesis-saving tinkering is allowed, to account for apparent planetary motions. But if one confines it to that role it is impossible to falsify it. The Copernican-Newtonian theory would be preferred to it only on the ground that it accounts for more, for example, the phases of Venus, the aberration of light and stellar parallaxes. As will be argued later, the possibility of falsifying a theory is as important as, if not more important than, the possibility of confirming it.

A more convincing case can be made for an instrumentalist view of theories of a more abstract and relational kind than those assuming terms such as genes and fundamental particles. What are known as scientific laws provide an especially suitable subject for the instrumentalist case. Newton's laws of motion and his law of gravitation will be used to examine the case.

In his book *Philosophiae naturalis principia mathematica* Newton developed his mechanics first in mathematical terms and subsequently gave a physical application. Thus his three laws of motion were stated as axioms in what he meant to be a mathematical system. Admittedly this system like Euclid's geometry contained terms which had physical referents. Nevertheless a case for the arbitrariness of his axioms (akin to the arbitrariness of Euclid's axiom about parallel lines) can be made in respect of the three laws. As stated by Newton (or more correctly as translated from his Latin), the first and second laws are:

Law I. Every body perseveres in its state of rest, or of uniform motion in a right line, unless it is compelled to change that state by forces impressed thereon.

Law II. The alteration of motion is ever proportional to the motive force impressed; and is made in the direction of the right line in which force is impressed.

Aristotelian physics had assumed terrestrial bodies 'in their own element' to be inert so that they would remain at rest unless pulled or pushed by some other body in motion. What continued to push a projectile after it left the hand was something of a puzzle. Buridan in the fourteenth century AD maintained that the external force imparted an impetus to the projectile which maintained it in motion until the work so involved expended the impetus. If the external force were imparted by an object moving on a curved path then the impelled object would continue on that curved path. Newton considered that continued motion was to be regarded in the same way as a continued state of rest, namely as an expression of the force of inertia 'by which every body . . . endeavours to persevere in its present state'. The further innovation which distinguishes this Newtonian principle from a step taken by Galileo in its direction was that if the body were in motion it would through inertia continue to move uniformly in a straight line. That is, Newton, following Descartes, regards 'instantaneous' motion as rectilinear, and curvilinear motion as the result of an impressed force operating to divert such rectilinear motion. It might seem that we can settle for Descartes and Newton and against Buridan by considering such cases as a stone released from a sling. The impetus imparted to it has come from the circular motion of the sling yet the stone does not continue to go around in a circle but shoots off in what looks at first like a tangent to that circle. However, when one follows its path further it is clear that it curves down towards the Earth. Thus neither Buridan nor Newton seems to be right. Newton, however, attributes the downward curve of the stone's flight to some impressed force exercised by the Earth and in the direction of the Earth. Careful consideration suggests that simple examples such as this can no more directly establish Newton's laws of motion than diagrams can directly establish Euclid's parallels axiom. Except in such cases as bumped billiard balls and a ball captive on an elastic cord, our evidence of an impressed force is, by the two laws, a change in the body at rest or in the body's uniform motion in a straight line. It would seem, then, that what we have in Newton's laws of motion are ways of analysing the motions of bodies so that we may think about them more effectively. It may be a matter of asking not: 'Is Euclid's geometry true or is Lobachevski's geometry true?', not: 'Is Buridan's law true or are Newton's laws true?' but instead which of the two alternatives is the more effective way of arguing about objects in motion. There remains, of course, the tantalizing question: 'Why are some ways of thinking about nature more effective than others if these ways are not founded on the way nature is in reality?'

When we come to Newton's physical application of his system, the instrumentalist interpretation seems to lose some of its cogency. Newton added to the general or mathematical laws of motion the law of gravitation which he stated as follows:

All bodies whatsoever are endowed with a principle of mutual gravitation. Every two bodies gravitate towards each other in proportion to their masses and in inverse proportion to their distance.

It is worth considering a little what can be meant by the words 'bodies gravitate towards each other'. Looking at the origin of 'gravitate' in the Latin word *gravare*, we may decide that the bodies weigh down towards each other. 'Mutual gravitation' suggests that their weighing towards each other may be deemed to be a tendency for them to move towards each other or for them to pull towards each other. Unless the bough is flexible enough to bend, the apple does not move towards the Earth before the stalk breaks. Nor for that matter does the Moon move towards the Earth in any literal sense; it moves on a path which is a compromise between the path attributable to inertia alone and the path attributable to gravitation alone. We may think of 'gravitate towards' as defining the motion of a body were nothing to restrain it or to divert it from that path, or we may think of it as any example of the 'force' we can feel through our muscles when we try to move stationary bodies, for example a fallen log across a road, or to resist certain bodies in motion, for example a cricket ball on its way to the boundary or a dog escaping from the discomfort of its weekly dip in the tub. If we adopt a meaning akin to 'force' as illustrated, then Newton's law in more precise form states that bodies exercise a mutual force (F) directed from one to the other in accordance with the equation:

$$F = G\, m_1\, m_2\, /\, r^2$$

This gravitational force referred to by Newton is not only that which we can observe kinaesthetically by hefting any object heavier than air or by holding down any object lighter than air and which we can see at work when the support is removed from an object which is heavier than air and at some appreciable distance from the solid surface of the Earth, for example when the stalk of an apple parts from the bough. It is also the force which stretches the spring in a weighing scale when an object is weighed by means of it. In relation to this later example, it must be said that we can establish observationally that a given object remaining invariant, as far as we can tell, will stretch the spring less when the weighing is done at the top of a high mountain than when it is done at sea level. That is, when the distance between the centre of the Earth and the object is greater, the force exercised on the spring is diminished. A series of observations will confirm or falsify that this diminution is in proportion to the square of the distance of the site from the centre of the Earth.

Newton considered that the gravitational force exerted by the Earth on other bodies such as apples might extend beyond the immediate terrestrial neighbourhood even as far as the Moon. According to the first law, the Moon would, unless deflected, move in a straight line forming a tangent with its actual or observed orbit. If the deflection from this straight-line path were the result of a gravitational force exerted by the Earth, then the angle of the orbit to its tangent would be accounted for by entering the masses of the two bodies and their distance from each other into the equation given above. Newton showed that the angle was deducible, within close limits, in this way.

The instrumentalist argues that this indicates not that the law of gravitation is a statement about anything in nature but that the law enables one to draw the right conclusions about certain observable situations. One instrumentalist has spoken of laws as inference licences or as rules for drawing certain sorts of inference. This, however, seems a confusion between the so-called formal sciences which are concerned with the structure and criteria of valid arguments and the so-called factual sciences which are concerned with observable situations involving terms and relations of a certain kind such as physical, biological, social and so on. The law of gravitation is not a rule for drawing inferences about moving bodies, it is one of the premises from which one may draw such inferences. When what is inferred is inconsistent with what is reliably observed, the scientist rejects the law, just as he would reject his assumption of some entity in similar circumstances. It could be, however, that it would not make much difference whether he is rejecting this as a useful rule of inference in drawing conclusions about what he observes in nature or as a conjecture about what indeed happens (unobserved) in nature.

What are termed working hypotheses may seem to lend some support to the instrumentalist view. These are hypotheses recognized to be inconsistent with some of the relevant observed facts but, in the absence of a tenable hypothesis, used because of their value in prompting further investigations. Nature seems to work in some respects *as if* what the working hypothesis asserts were the case and in other respects *as if* it were not. The trouble is that one cannot tell in advance where the working hypothesis will work and where it will not. Unless nature always works *as if* the theory were true, the theory has only wayward predictive value and is a very uncertain guide to the way in which one should think about those parts of nature one has not been able to explore through observation. Where nature does, as far as we can tell, always work *as if* the theory were true, there seems little point in denying that the theory is contending something about nature. To say that nature does not work in this way but only as if it did is to make a distinction which makes no difference.

There are, of course, stepping points in an argument and symbolic elements in the expression of a theory for which there may be no empirical referent. There may indeed be forty-three children in nine families in some community, but there may be nothing in that community that has 43/9 ($=4.\overset{.}{7}$) children in it. Yet this 'construct' or 'fiction' which we call an average may be very useful for the comparison of the trend in family sizes in this community with the trend in some other. In interpreting the verbal statement 'he learned to swim' we may be able to find separate referents for 'he', for 'learned' and for 'to swim' but not separate referents for 'to' and for 'swim'. As we have already seen, a purely dispositional concept anchored fore and aft may be treated simply as a convenient perch on which an elaborate flight of argument momentarily rests.

Varieties of explanation

The word 'explanation' derives from the Latin verb *explanare*, to spread out, to make clear. We use it in several ways. One usage refers to giving the meaning of words or of the concepts signified by words. Thus we explain 'mammal' by saying that it refers to an animal which suckles its young. This is sometimes said to be providing better understood words which are equivalent to less well understood words. Though the process is carried out through the medium of words, it is more concerned with what the words signify than with them as words in themselves. One would be more concerned with the word 'mammal' as such, were one simply to say, in analogy with the opening sentence in this section, the word 'mammal' derives from the Latin noun, *mamma*, a breast, a teat.

An extension of the usage just examined is the teasing out of the full significance of complex concepts. We all understand in at least a superficial way what is meant by 'motive' in statements like 'his motive for resigning his office was pique'. A careful analysis is required to reveal whether it means 'cause' or 'reason' or both or something else. Such an analysis may be termed an explication or an unfolding of the concept.

There are three other usages which would scarcely be anticipated from a consideration of those already given or even, perhaps, of the

Latin verb *explanare*. All make reference to the reason for some situation. A situation is explained when a, if not *the*, reason for it is given or grasped. The differences between the usages hinge on the separate ways the notion 'reason' may be taken. One may justifiably speak of a rationalist way and an empiricist way of taking it. The rationalist regards a situation as being explained (or its reason being apprehended) when it is appreciated that the situation could not, in any circumstances, have been other than it is. This appreciation may consist in recognizing that the situation is perfectly apt or right. It was this kind of insight which Kepler felt he had attained in his demonstration that the five regular solids fitted the five spaces between the Sun's planets known to him. Or it may consist in recognizing that the situation, though *prima facie* contingent, is required to be as it is by some indubitable principles.

The empiricist regards a situation as explained when he can show that it is implied from some supportable premises, ordinarily of a general sort. This usage has some logical features in common with the second rationalist usage, namely the implication of the situation by premises which seemingly refer to other matters. The difference lies in the status accorded these premises. The rationalist asks that they be necessary, self-evident, indubitable or in some other way unshakeable. The empiricist asks merely that they be testable. Thus the apparent motions of the 'wanderers' were explained by Eudoxos when he deduced them from the assumed motions of his nests of spheres, by Ptolemy when he deduced them from his system of deferents and epicycles focused on the Earth and by Kepler when he deduced them from his system of ellipses focused (except in the case of the Moon) on the Sun. The bases of explanation (the *explanans*) in all these cases imply what is to be explained (the *explanandum*). If the *explanans* is regarded as being inevitably or necessarily as it is stated to be, the approach is rationalist. For instance, Eudoxos and Ptolemy considered that the real motions of the planets *had* to be circular in figure and uniform in angular velocity. Though Kepler flirted with the thought of egg-shaped figures (because the egg is the beginning of life), he finally resorted to conic sections as a matter of brute fact and to this extent turned, no doubt reluctantly, from rationalism to empiricism.

The other rationalist mode of explanation stresses insight into the necessity or rightness of a state of affairs but does not include reference to the implication of the *explanandum* by the *explanans*. The necessity or rightness is transferred from the *explanans*, as the rationalist conceives it in the first usage, to the *explanandum*. It is the latter which is seen to be necessary or to be fit and proper. Such a view may be called intuitionism. In the history of science it has had many manifestations such as the teleology of Aristotelian physics, as vitalism in nineteenth-century biology and, more recently, as personalism in its idiographic form in psychology. The arguments advanced earlier against self-evidence as a base for rationalism have equal force against the insight of intuitionism. There is, for example, no way within the framework of the view whereby one's choice of one intuition can be defended against some contrary intuition; again, there is no way within the framework of the view whereby one can distinguish between a true intuition and a false one. The intuition may bring understanding (or *Verstehen* as its German proponents say) but this is an inner enlightenment rather than an elucidation of the grounds of a situation or of conditions under which it occurs. It is mysticism rather than science.

Reductive explanation

Granted that a scientific explanation, as our several case studies suggest, involves an implicative relation between the explaining propositions (laws, contentions about the arrangement of entities, etc.), the *explanans*, and the situation to be explained, the *explanandum*, we have to ask whether the relation is sufficient to constitute an explanation or whether there are some further criteria to be satisfied by the set of premises in such a relation before it qualifies as an *explanans*. There are some premises from which a situation may be deduced though they hardly constitute an explanation of it. If one wishes to explain the presence of a goat in one's garden, it will hardly do to say that this may be inferred from the presence of two goats there (even if indeed one has seen both goats there). Likewise it is hardly an explanation of a given negro's dark skin to say that all negroes have dark skins, though the deduction contained in the following is valid:

> All negroes are dark-skinned.
>
> The Emperor Jones is a negro.
>
> Therefore, the Emperor Jones is dark-skinned.

Perhaps these premises are too 'near' to the situation which has to be explained and are not 'general' enough. The rationalist would say that these premises are themselves contingent whereas *explanantes* need to be necessary. The empiricist has no such ready answer though he should recognize the need for one. He may make reference to laws, to propositions of greater generality and greater degrees of abstractness but he may also find himself in difficulties in saying precisely what these contentions mean. He may indeed be driven to speak of the sciences as forming an inverted hierarchy and of explanation in one science as the reduction of its phenomena to a 'lower level' science which because its laws are more simple is more secure than the phenomena of the 'higher level' science.

We have already encountered the view that the sciences lie in a series of strata or levels, with physics (if one confines the series to the factual sciences) at the base, and with chemistry, biochemistry, biology, psychology, and the social sciences each at successively higher levels. We have already seen a case where some factual data of chemistry, arranged in a clear pattern by Mendeleyev and his successors, has been explained in terms of a physical theory introduced by Rutherford and Bohr for strictly physical purposes. We have also seen strong hints that the factual data of genetics in biology may become explicable in terms of a biochemical hypothesis about the structure of the DNA molecule. The question for us is whether these are merely some of the explanations to be encountered in science or whether they are the paradigms of all general explanations.

There is a certain plausibility in the view that the subject-matters of the sciences fall into an order like that propounded in the hierarchical view. Many physicists can proceed effectively with their work without paying much heed to chemistry and any heed whatsoever to biochemistry, biology and psychology. However, few chemists, if any, can ignore the teachings of physics. Indeed one may say that chemistry has been shown to be an extension of physics. Likewise biochemistry is an extension of chemistry. Increasingly the biologist concerned with the whole organism is obliged in his search for explanations to go beyond cellular biology to molecular biology, that is, to biochemical processes in the cell as well as to the more molar and traditionally biological features and functions of the cell. The psychologist in trying to set himself up as an experimental scientist rather than as a special kind of philosopher has usually seen the need to relate his psychological findings to neurophysiology, to endocrinology and other aspects of chemical physiology, and to genetics. There seems then to be little doubt of the felt order of dependency.

169

There was, however, a time when it was argued that there were fundamental discontinuities between the sciences. In particular, it was urged that there were impassable chasms like the Wallace Lines which divide regions with distinctive types of flora and fauna—one between the sciences dealing with inorganic matter and those dealing with living processes and another between those dealing with the bodily processes of living organisms and those dealing with their mental processes. A further break was often suggested between the behaviour of individuals and the processes of society as a supra-individual organism. An example of the last of these three is to be found in the Great Man versus the Social Movement (or Spirit of the Times) views of history. One suspects that some who urged these Wallace Lines between the sciences did so in order to defend the existence in its own right of the 'higher' science in which they were interested. The recognition of this protective motivation, however, does not establish that the separatist case is wrong.

Vitalism, the doctrine arguing for some distinctively vital principle infusing living matter, recognized that the physico-chemical processes to be found in living organisms formed a substrate upon which life processes occurred. However, it maintained that the properties and processes of life, even if occurring upon an inanimate base, were not deducible from or derived from the laws of interaction of inanimate matter. The inorganic seemed manifestly incapable of growth, adaptation and reproduction which are characteristic of all organisms, and of sensitivity, spontaneous motility and learning which are characteristic of animals. Hence, these two sets of features were deemed to be something superimposed, the first upon inanimate nature and the second upon the vegetative aspect of animate nature.

Mind-body dualism argued for a comparable break or series of breaks between the bodily processes and the mental life of animals, especially the vertebrates. Perhaps a first break was to be defined in terms of purpose or the pursuit of ends distinguishing animals from plants and deriving from this or co-ordinated with this, some simple judgemental processes going beyond mere sensitivity and discrimination. A second, if not clearer, break seemed to be that which cut off self-determination (or free will) and the co-ordinate powers of abstraction, conceptualization and symbolism. All these powers seemed basic to the moral, intellectual life characteristic of, if not peculiar to, man.

Vitalism and mind-body dualism are not popular in present day biology and psychology though there are still occasional vestiges of dualism in psychology. The pendulum has swung away from such separatist views towards reductionist views in which living processes are regarded as nothing but complex physico-chemical activities and mental events as nothing but complex bodily processes. In saying 'nothing but' it is usually maintained that there is nothing in the 'higher level' science which is not explainable in terms of, or not deducible from, the 'lower level' science. The advances of biochemistry have produced previously unexpected links between the chemical and the biological by showing that growth, reproduction and adaptation rest upon complex chemical processes. Likewise the advances of neurophysiology and endocrinology have shown that many features of behaviour are affected by 'lower level' processes. It is important, however, to distinguish between strict or literal reductionism and a recognition that events at a 'higher level' are conditioned by those at a 'lower level'.

In between the extremes of separatism and reductionism is a view often termed emergentism. It maintains, for instance, that though life is based on the inanimate, it manifests properties which are not to be found in the separate inanimate components but which emerge only when those components are arranged in certain ways. These biological properties are deemed to be peculiar to the whole (the cell or the organ or the organism as the case may be) and may be discovered only through the study of that whole. It is argued that not only are these properties of the whole not present in the components but also that they are not deducible from propositions about the properties of the components. Emergentism does not confine such emergence of whole-qualities to the major 'breaks' detected by separatist doctrines between the inanimate and the living, between the body and the mind, and between the individual and society. It considers that the distinction in chemistry between mixtures and compounds illustrates its principal tenet. One may place two volumes of hydrogen and one of oxygen in a jar and simply have a mixture of the two gases and a summation of both sets of properties. On passing an electric current through the mixture, the atoms of the two gases combine to form water. This compound has properties different from, even at variance with those of its 'parts' when they were separate. Thus water may be used to extinguish fire whereas hydrogen is highly inflammable and oxygen supports fire. Similarly it is said that though a living cell can be broken

171

down into molecules of greater or less complexity—proteins, fats, carbohydrates, mineral salts, etc.—its distinctive properties are neither to be found in these molecules nor to be deduced from their properties. The argument for emergentism in simple cases such as water and other chemical compounds has been seriously undermined by the theory of sub-atomic structure developed by Rutherford, Bohr, and others. While it is still true to say that water has properties not present in its components, hydrogen and oxygen, when separate, it is no longer true to say that the properties of water cannot be deduced—not, of course, from a consideration of the properties of hydrogen and oxygen alone but from a consideration of their properties in conjunction with the theory. The relation of life to physico-chemical processes is still a matter for dispute, though the reductionist argument has over the last half-century been gaining more and more evidential support. Some biological events may already be deducible from 'lower level' considerations, but in most cases the situation is as we found it to be in genetics—the promise rather than the attainment of reduction has been obtained. The situation at the 'break' between the mental (or the behavioural) and the biological is more obscure, although there too the separatist has continually lost ground as evidence of the conditioning of mental (behavioural) processes by bodily processes has accumulated.

The argument between reductionism and emergentism may be clarified by briefly considering an example of reduction which occurred in physics in the late nineteenth century. Classical mechanics (the physics of particles in motion) as developed by Galileo and Newton and thermodynamics (the physics of heat) grew up as separate sciences in the sense that though they had some common terms and laws, they each had some distinctive terms and laws. For instance, one could deduce Kepler's laws of planetary motion from Newton's mechanics including his law of gravitation when applied to given situations but one could not deduce the gas laws of Charles and of Boyle from these premises or any other premises to be found in classical mechanics. However, when some additional propositions were introduced into mechanics, it became possible to deduce the whole of the formerly distinct science of thermodynamics from this supplemented mechanics. Thus thermodynamics was said to be reduced to mechanics. The additional propositions needed for this reduction included an hypothesis about the molecular constitution of a gas, a statistical assumption about the motions of these molecules and a definition of temperature in terms of the kinetic energy of molecules. None of these formed a part of what was originally mechanics, though none was at variance with anything in that science. Strictly speaking thermodynamics was reduced to a supplemented mechanics.

One may suspect that similar supplementations will be required in the case of other reductions of one science to another. These supplementations are of at least three possible sorts. First, the definition of terms in the 'higher level' science by means of terms in the 'lower level' science. The identification of a gene as a 'run of treads' in the staircase of the DNA molecule may be an example. It is important to note that it does not subtract anything from what gene originally meant, and, of course, it adds something to what one might know of the DNA molecule on purely chemical considerations. Second, supplementary laws or hypotheses added to those already promulgated in the 'lower level' science. The statistical assumption concerning the motions of molecules in the reduction of thermodynamics may be an example. One may reasonably graft on to classical mechanics such a 'statistical law' even though the need for it had not been felt prior to the incorporation of thermodynamics into mechanics. Third, there are hypotheses in the 'higher level' science formulated to provide a foundation on its bank for the bridge which is constructed to link it with the 'lower level' science. The hypothesis that gases have a molecular constitution, introduced to enable the reduction of thermodynamics, may be an example. Just what supplementations of biology are needed for the reduction of psychology are at the moment largely a matter of vague guess-work.

The reductionist and the emergentist agree upon the unity of the sciences, a view at variance with the separatism of vitalists, dualists, and others. Their dispute is about whether the 'higher level' sciences have to consider anything more than the 'lower level' sciences. It is being suggested here—the matter is so difficult and uncertain, that anything beyond a suggestion would scarcely be justified—that something more will have to be introduced into biochemistry before biology can be reduced to or deduced from it, and that something more will have to be introduced into biology before it can be an adequate reduction base for psychology. That is, it is suggested that, as in the case of thermodynamics, the reduction will not be complete. Thermodynamics is not a logical extension of classical mechanics in the sense that the latter provides all the premises needed for the deduction of the former. However, we may say that something more than the mere unification of thermodynamics and mechanics occurred in that the latter provided the more 'basic' material in the relation of the two. One may reasonably expect a similar situation to emerge in the case of physics, chemistry and biochemistry providing a basis of explanation for biological phenomena. The emergentist feels, perhaps wrongly, that 'higher level' phenomena are somehow undermined or eroded away by suggesting that they may be reduced, even in the moderate sense used above, to 'lower level' laws and that the reductionist wrongly claims too much. Whatever one's hunches about the matter, the answer remains for the future to give.

A few additional remarks will be in place before the topic is left. The 'lower level' scientists tend to make their task easier by excluding 'higher level' considerations. A social scientist is expected to take his phenomena as they come, whereas the physical scientist picks and chooses. For instance, suppose we get the latter to profess that his task is to account for the occurrence of physical (i.e. non-living) phenomena, and then we ask him to account for a particular bridge across a stream. He will be inclined to ignore such issues as why the bridge is at this point rather than at some other, why it is a suspension rather than an arch or cantilever bridge, and so on, on the ground that these are the product of various human and social activities which are not his concern. A social scientist would be accused of quite unjustified separatism if he ruled out all physical considerations in accounting for a social phenomenon whether it be stove-pipe pants, the Tower Bridge, the Australian accent or the antipathy that grew up after World War II in Great Britain to persons of negro race. Possibly the 'higher level' sciences cannot afford to be a-historical whereas the 'lower level' sciences have thriven on dealing with the 'pure case' without concern for its historic setting. Reductionism may be an appropriate philosophy of explanation when one is dealing with 'pure cases', whereas explanation from any suitable premises, 'higher' or 'lower' may be appropriate when one is dealing with events identified with specific space-time locations. In so far as the factual sciences use the formal sciences as models, they are likely to adopt at least a moderate reductionism. In so far as the 'higher' members of the family depend upon commonsense considerations they are likely to be emergentist if not outright separatist. Separatism, however, would seem to be doomed in the long run. The growing evidence of inter-relation between the subject-matters of the sciences argues for an increasing unification of the sciences even if it does not take the form of the reduction of the 'higher' to the 'lower'.

What may serve as explanantes?

Several views about what may serve as the basis of explanation in science have already been considered. Some of these have been argued against in a way which has called for their rejection, others only to an extent calling for caution. The rationalist view that explanations in science must be brought back ultimately to some fundamental, unassailable truths different altogether from the questionable and partial proposi-tions about observed and observable situations has been rejected here. It has been said that there is no coherent way of saying how any proposition about what obtains in nature can be altogether different from those propositions expressing our observations of nature. Further, it has been said that there is no justification for regarding any proposi-tions about what obtains in nature as unassailable, as necessarily true. There may be some *a priori* truths, but those for which the strongest claim can be made seem, if this claim can be sustained, to have nothing to say about what obtains in nature. The simple view, supported, if by anyone, only by those swinging back too far from rationalism, that any proposition which implies another provides an explanation of it has been shown to be inadequate. There are many simple logical transformations in terms of implicative relations which have none of the main functions of a scientific explanation.

A view that has proved more tempting to many recent scientists and philosophers of science is reductionism—the explanation of an event in one science is provided when it can be shown to be implied by a more 'fundamental' science. Certainly many explanations conform to this pattern: two examples are provided by our case studies. How-ever, some doubts have been thrown here on the hierarchical ordering of the sciences and on the tacit assumption in reductionism that while everything in, say, biology may be accounted for in terms of physical laws and events, nothing in physics can be accounted for in terms of biological laws and events. There are many physical phenomena for which we cannot fully account without reference to biological events. One may mention coral atolls, coal measures and oil basins. Though these phenomena may be ruled out of the realm of physics proper, they are certainly within the realm of the physical sciences.

Reductionism and the notion of a hierarchy of sciences provide a clue to the criteria of a genuine explanation. When a reductionist explanation is offered resort is had to propositions more general than those in which the *explicanda* are asserted in the sense that they have a wider range of reference or application. This greater generality of reference is achieved through a higher degree of abstractness, that is less reference to the specific combination of properties in particular cases. What seems to have made the rationalist's case plausible is that the most impressive explanations are those in which some of the bases of explanation are of great generality and are so abstract that they can be seen to have reference to the particular events to be explained only after long chains of inference are unfolded. Scientific laws have these features, physical laws having the possibility of greater generality in this sense than biological laws, and biological laws than psychological laws, and so on. However, inferential chains do not necessarily run from physics to biology, from biology to psychology and so on, even though especially strong ones do take this direction. Further, laws alone are not enough for the deduction of particular events. Other and more specific considerations need to be taken in conjunction with them. We cannot work out the area of a rectangle if all we have is the mensurational rule:

Area = length times breadth.

We must have the specific values for length and breadth. It could be that this is the way in which biological, psychological, and sociological considerations enter into the explanation of physical phenomena.

EPILOGUE

Though scientists expend a great deal of their collective time and effort in the development and testing of theories, and though the member of their company who produces an important theory is more likely to be remembered than the one who discovers an important fact, facts form the base to which they all continually return in their pursuit of knowledge. In this they differ from the creators of formal systems in mathematics and from the proponents of metaphysical schemes. The mathematician generates his system by deducing theorems from the initial axioms and definitions. Though some added credibility may be conferred on the premises through reference to situations encountered in nature, the main justification of the axioms is that they are a minimum set which will generate an internally consistent set of conclusions. Likewise the metaphysician tries to enlighten and to justify his more specific contentions, which may have greater or less bearing on the observable world, by arguing that they follow from some fundamental principles which he deems to be generally acceptable if not unassailable. The more the metaphysician is a rationalist, the more likely is he to deem his principles *a priori* truths or to be self-evident or able in some other way to stand on their own feet. The more he is an empiricist, the more likely is he to deem his leading propositions to be supported by commonsense or to be derived from common experience with simple or 'pure' situations. But however they are considered to have been laid, these general propositions are the foundations which have to bear the weight of the rest of the metaphysical structure.

The scientist, or more often the philosopher of science, recognizing that theories cannot be deduced from facts as theorems in pure mathematics are deduced from axioms, has often been tempted to assert an alternative mode of inference. However, as has been pointed out above, none of the attempts to demonstrate that there is a valid inductive mode of inference, whereby theories are drawn from facts, has been successful. Subject to a reservation to be stated later, the scientist does not infer his theories from his facts in any logical sense of the word 'infer'. He invents or imagines or constructs his theories. It is true that he is assisted by many clues and suggestions from his facts. He often works through analogies with other better-known or better-understood situations. But it can never be maintained that 'as the facts are such and such, then it follows that this theory is correct' as it can be maintained that 'as the axioms and definitions are such and such, then it follows that this theorem is correct'.

The reservation mentioned above concerns the way in which the scientist may be able to fix some of the details of his theory by valid inference from premises which include some of his facts. Theories consist of both general and specific materials. For instance, the sets of deferents and epicycles focused on the Earth constitute a general component of the Ptolemaic theory. The ordinary diagrammatic representation of the theory shows little more than this. But for this general component to work a number of details must be introduced such as the periods of deferent and of epicycle and the ratios of their radii. These details may be determined by a guess-and-try procedure or by a process of working back from the facts taken within the context of the general components of the theory. In some circumstances, valid inference is involved in the latter process but it is deduction and not induction. In any case it is not what is in mind when the possibility of inducing general laws from particular facts is being considered.

The scientist invents or constructs his theories for several purposes. He may, as Harvey did, do so to fill gaps in the facts available to him. He may, as Mendeleyev did, do so in the effort to show an order or pattern amongst facts that would otherwise seem to be disorganized. He may, as Newton did, do so in order to provide an explanation of some set of facts ordered by the theorizing of others. Strictly speaking it does not matter much what the scientist's purpose was; it is the outcome that matters. Bohr did not set out to explain the periodic table. This does not affect the soundness of the contention that his theory of sub-atomic structure does explain the regularity revealed in the table. Further, the value of the theory is enhanced by this additional success.

It is the explanatory role of theories which provides us with the main inferential relation between theories and facts, the inferences being drawn from the theories. In respect of the logic of the situation the theories of empirical science are akin to the axioms in formal systems and the facts which the theories explain are akin to the theorems in such systems. In other respects the roles are quite different, if not the converse of each other. In empirical science the facts are deemed to be the base of operations, the foundation on which the whole structure rests, whereas in formal systems the axioms form the base or foundation. The scientist places this emphasis on the facts partly because he regards them as being more reliable but mainly because the only scientific point of theories is the service they perform in respect of facts—rounding the available facts out into a coherent picture, suggesting a pattern or order amongst them or explaining them or anticipating and so providing leads to the observation of further facts. Without the facts the theories would be pointless—in certain circumstances, of course, a fact with no theoretical significance is regarded as trivial. In the creative arts, an imaginary construction, such as a poem, a song or a picture, may have aesthetic value even though it has no reference or relevance to the everyday world around us. But scientific theories have no value as such unless they have reference to the observed and observable world. Some far-ranging theories of a metaphysical kind which scientists as well as philosophers propound may have more aesthetic than scientific value. But if they have no relevance to facts, more specifically if they are not testable through facts, they lack any scientific value whatsoever.

Before we proceed to summarize the way theories are tested by facts, it will be desirable to consider the reliability of facts. It is only when one encounters facts which are at odds with one another that one recognizes the need to question their reliability. Otherwise one takes them without question. 'I tell you it is there, because I just saw it there' is ordinarily regarded as an overwhelming reply to someone who denies the presence of an object in some place. But if the objector is not overwhelmed by it he should be when we lead him to the site and point the object out to him. Concurrence between two or more observers is deemed to be good evidence of the reliability of what they separately report. There are, however, exceptions. An extreme case is provided by 'mass' hallucinations. A simpler and clearer case is provided by geometrical illusions. Consider the design shown in Fig. 9.1. To the unaided eye the pair of lines passing from left to right seem bowed when the design is viewed front-on. Any number of independent observers will agree on this. However when a straight-edge is laid along these lines or when they are viewed by sighting along them when the figure is viewed side-on, they appear straight. Our several observers will also agree on this. For various reasons, the latter judgement is regarded as more reliable than the former. Important amongst these is the comparison of the lines with an object which in

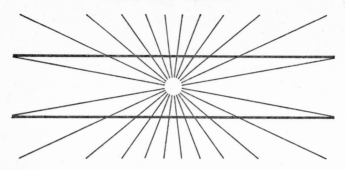

Fig. 9.1   The Hering Illusion. The pair of transverse lines are straight. The apparent bowing is an illusory effect produced by the radial lines.

other situations is demonstrably straight and which seems not to have undergone a change in this situation. Further evidence can be obtained by masking the radial lines without obscuring the horizontal lines.

The point to be made is that the evidence of our senses is fallible. The facts of observation may be false. Such an admission should not drive the scientist into extreme scepticism, the doctrine that as one cannot be sure about everything one observes, one should not believe anything. It leads him or should lead him to be critical, to ask for the credentials, as it were, of anything claimed to be a fact of observation. This involves attempting to establish the reliability of the alleged fact. Concensus, congruence, and coherence are important but not infallible criteria of the reliability of facts. If two or more observers agree upon what is observed independently by them, the fact reported by them is deemed more reliable than would be the case if they disagreed or if they agreed in a situation where one could have influenced the other. If several modes of observing a situation lead to the same reported fact, the fact is deemed to have enhanced reliability. If the fact fits in with other facts or with theories well supported by other facts, it is deemed to have greater reliability than it would were it an isolated fact. However, none of these criteria are absolute. If they were, there would only be one sort of revolution in science, namely that which involves a radical change in thought about an existing set of facts. The Copernican and Darwinian revolutions were mainly of this sort. There have, however, also been revolutions, of which the Einsteinian is one, sparked by incongruent facts. Though these criteria are not absolute, they guide the scientist in his fact-finding. Improvements in the techniques of observing are introduced in order to lessen the range of variation in repeated observations, that is, in order to reduce variable or random error. Observations are repeated under varied conditions in order to eliminate or to reveal systematic error, that is, obscuring or distorting influences from extraneous sources. This control of conditions such that some are varied and some kept constant is the distinctive feature of the experimental method in science. It may be achieved in several ways. The conditions themselves may be manipulated by the experimenter or he may select appropriate naturally occurring cases or he may disentangle by mathematical means the several possible influences at work.

The testing of theories by means of facts may be direct or indirect. Harvey assumed that there are capillaries linking the arteries and the veins. Malpighi looked for them and found them, so confirming Harvey's theory. Had he not found them the theory would have been 'not confirmed' or as some like to put it, 'infirmed'. Copernicus assumed that the Earth is in motion around the Sun but this is not something any earthbound observer can observe. However, Copernicus' assumption implies stellar parallaxes, phenomena which may be observed if they occur with a magnitude detectable with the technical means available to the observer. Tycho Brahe could not detect any stellar parallaxes and regarded this failure as infirming the heliofocal theory; nor could Bradley, though he seems to have suspected that his failure was the result of the limitation of his equipment. Bessel, Struve, and Henderson each observed stellar parallaxes in the late 1830s; their observations were deemed to be confirmations of Copernicus' theory. Malpighi's confirmation of Harvey was direct, whereas Bessel's, Struve's, and Henderson's confirmations of Copernicus were indirect. Malpighi's observation corroborated Harvey's assumption, whereas these astronomers' observations corroborated an implication of Copernicus' assumption and not any explicit Copernican assumption.

At this point it will be useful to distinguish between the notions 'confirm' and 'verify', and between 'infirm' and 'falsify'. For the purpose of the distinction let us assume initially that the testing facts are completely reliable. When a direct test confirms a theory, it shows it to be true, that is the fact literally verifies the theory. When an indirect test confirms a theory, the fact does not literally verify the theory. The fact in this case does no more than show that the theory may be or could be true. The observed positions in the sky of the several planets confirm the theories of Eudoxos, of Ptolemy, and of Copernicus equally (if we set aside an increased precision as we move from earlier to later theories, a change to be attributed mainly to refinement of the details of the theories rather than to their central notions). In general terms these facts show that any of the three theories *could be* true but they do not show which, if any, *is* true. The observation of the phases of Venus infirmed the Eudoxan and the Ptolemaic theories and confirmed the Copernican. That this fact also confirmed the Tychonic theory indicated that it did not verify the Copernican. The observation of stellar parallaxes infirmed the Eudoxan, the Ptolemaic, and the Tychonic theories and confirmed the Copernican, but as it would confirm any theory which has the Earth in motion its confirmation is not exclusive to the Copernican theory.

Granted that the testing fact is completely reliable, it falsifies the theory when at variance with it whether the test be direct or indirect. Had Malpighi been able to report with complete reliability that there were no small channels linking the arteries and veins, then Harvey's theory would have been falsified. Likewise, had Tycho Brahe's report of no stellar parallaxes been a completely reliable fact, the Copernican theory would have been falsified. It is important, however, to distinguish between positive, negative and null factual evidence. The observation of stellar parallaxes is positive for a heliofocal theory and negative for a geofocal theory. A failure to observe stellar parallaxes (as distinct from an observation that there is none) is a null finding. It is more difficult to establish the reliability of null observations than of positive or negative observations. Consider the case of phases of Venus as implied by the Copernican theory, namely a series running from 'new Venus' at inferior conjunction in the middle of the retrograde phase of apparent motion, through crescent, half, gibbous to 'full Venus' at superior conjunction coming back through gibbous, half, waning to 'new Venus' again. A series of observations corroborating this prediction is positive in respect of the Copernican theory. A finding of any other pattern, for example that which was implied by the Ptolemaic theory when taken realistically, would be a negative finding for the Copernican theory. A report that the form of the planet could not be seen clearly enough to reveal whether there were phases or not would be what is meant here by a null finding. A report which could be taken to show that Venus has no phases would constitute a negative finding in that it would contradict the implications of both the theories we have considered. It would be proper, however, to ask whether such a report should be so interpreted or whether the observations have to be regarded as unsuccessfully carried out. Mercury, Venus, Mars, Jupiter, and Saturn are seen by means of even a modest telescope to present a disc-like appearance as the Sun and the Moon do to the naked eye. Even the largest telescope fails to show bright stars such as Sirius, Canopus, and Betelgeuse as anything but points. We have many other grounds for believing that these distant stars have much greater diameters than any of the Sun's companions, but because of their distance and the limitations of our observing devices we would not be able to see them as going through phases if indeed they did so. Again, we may observe not only the sunshine on the Moon but also, in certain circumstances, the earthshine on it. We may suppose that Jupiter's companions have some reflected Jovian glory, but as we can see them and photograph them virtually as little more than 'points' (and not as clear discs) we cannot see anything but their reflected sunshine. The failure to see the stars as discs or to see the 'thorshine' on the Medicean planets are null and not negative findings. However, if we did not have other evidence from independent theoretical considerations to support us in this view, we might be tempted to regard these failures to observe particular situations as constituting negative rather than null findings.

The infirming facts in direct tests of theories have usually to be regarded as null. They are difficult to interpret and, as a consequence, difficult to assess in respect of reliability. Thus it is customarily said that they do not falsify the theory. In these circumstances, in the direct test the theory is either verified (granted the facts are highly reliable) or infirmed (granted the facts are ambiguous or of dubious reliability). In the indirect test, positive facts, no matter how reliable, can only confirm the theory; reliable negative facts on the other hand falsify it. It is often said that complete confirmation is verification and complete infirmation is falsification. In that case, in the direct test, theories may be completely confirmed (i.e. verified) but, in general, only incompletely infirmed, whereas in the indirect test, theories may be completely infirmed (i.e. falsified) but only incompletely confirmed. This statement depends, of course, on the assumption that positive and negative observations may be treated as completely reliable in certain circumstances whereas null observations may never be so treated.

There is another kind of null observation. The kind we have considered is unsuccessful in that it does not establish what obtains in nature. The other kind establishes what obtains in nature but not in a way that discriminates between theories. To show that blood consists of corpuscles borne in a clear fluid may be a well-established fact but it is null or neutral in respect of Galen's and Harvey's theories. That is, it is a fact compatible with both of them but required by neither.

A theory designed to fill gaps in the available facts, as Harvey's was, is open to the direct test. Some assumptions of entities such as genes introduced for explanatory purposes may also be open to the direct test if the assumed entities become identified with some observable properties or situations. Usually, however, explanatory theories are subject only to the indirect test. It is important in such cases that the theory imply observable situations other than those it was introduced to explain. Otherwise if it were well designed then the test can do nothing but confirm it. The test in this case is no more than a checking operation like that taught to the schoolboy as a means of checking the accuracy of his calculations. The recognition of the need for an unprejudiced test of a theory reveals a weakness in the instrumentalist view of theories. If we regard the Ptolemaic theory as a fiction convenient for the calculation of planetary positions at given times, then it does not matter greatly that it will not let us deduce the relative constancy in the angular size of the Moon or the precise changes of apparent brightness of the naked eye planets or the telescopic phases of Mercury and Venus. That the amended version of the Copernican theory does enable these deductions is regarded by the instrumentalist as evidence merely of the greater convenience or range of application of that theory. But if we regard both theories as conjectures about what obtains in nature, then we cannot shrug off the misleading implications of the Ptolemaic theory for other than positional data. We may conclude not only that the geofocal theory is less convenient but also that it has been shown to be false. Knowledge may be gained by the elimination of both ignorance, a null state of affairs, and error, a negative state of affairs. The instrumentalist view of theory minimizes the importance of the latter means of attaining knowledge. In the context of that view, useless theories can be shaken off—a result that should not be unappreciated—but false theories which have some uses can be tolerated, not merely as makeshifts but as respectable theories of narrow application. The realist view of theories requires that all the implications be followed up, not merely to see whether the theory has wider application than its inventor meant it to have, but more importantly to expose it to possible falsification.

In the discussion above of confirmation, verification, etc. it was taken for granted for the moment that the factual material was completely reliable. On many occasions, the scientist too must regard it so, at least for the time being. In an earlier chapter (*see* p. 154) reference was made to walking on solid ground and skating on thin ice. It is important for the scientist to keep on asking which is which, but there is no way of his being absolutely certain in advance of taking a step. Should he become obsessive, should he doubt every possible stepping place, he will take no steps at all. He must, however, be ready to decide on whatever evidence he can summon that this spot rather than some other is the one likely to bear his weight—and to act on that decision. Sometimes he will get a ducking, but with a bit of shrewdness, a modicum of other knowledge, sound techniques and some luck, he will also make progress in achieving knowledge of the world of which he is a part.

An interesting observation is that the scientist to be successful does not need to have a clear explicit understanding of the grand strategy that he seems to use. In many cases he has little more insight than the caterpillar has into the neuro-muscular physiology of his walking. Many scientists who have tried to formulate the logic of science hold quite erroneous views, granted that the foregoing analysis is sound. Most do not bother to formulate it in any detailed way. There may seem to be a paradox here. It would be nice to think that a study of the philosophy of science would make good scientists better ones. But the brute fact is that the caterpillar does not need to be a neuro-physiologist, the poet a literary critic nor the scientist a philosopher of science. There is an element, greater or less, of art in all their undertakings. There is something besides both critical understanding and art. The complex beauty of sub-atomic events, the fearful symmetry of planetary motions and the transmission of hereditary characteristics through the marvellous coding in the DNA molecule involve, except in the view of theologians and teleologists, neither insight no art. They all work in these economic, effective and seemingly inexorabl ways. The scientist, though a student of nature, is also part of nature His methods whether he understands them or not are often economi and effective; when they are, there also seems to be some quality o inexorability about them. Like any other event in nature they ar worthy of study. An attempt has been made here to show in a tentativ way what features they have when they are economic and effectiv and what features result in their being something less.

Further Reading

Though the four case studies were introduced here in order to draw attention to certain problems in the philosophy of science, they are inherently interesting. Material suitable for the general reader is in short supply on the first and the third. There is an embarrassment of riches on the second and an adequate supply on the fourth.

THE MOTION OF THE BLOOD

BAINTON, R. H., *Hunted Heretic: The Life and death of Michael Servetus, 1511-1553*, The Beacon Press, Boston 1953.
A moving biography of Servetus.

FLEMING, D., 'Galen on the motions of the blood in the heart and lungs', *Isis*, Vol.46 (1955), pp.14-21.
Calls Singer's account of Galen into question.

ROOK, A. (ed.), *The Origins and Growth of Biology*, Penguin Books, London 1964.
Contains extracts from Harvey and Malpighi.

SINGER, C., *A History of Biology*, rev. edn, Abelard-Schuman, London 1960.
Material on Galen, Vesalius, Servetus, Fabricius, Harvey, and Malpighi is to be found in the early chapters.

THE MOTIONS OF THE PLANETS

COHEN, I. B., *The Birth of a New Physics*, Anchor Books, New York 1960.
Discusses the switch from Ptolemaic to Copernican-Newtonian conceptions.

DREYER, J. L. E., *History of the Planetary Systems from Thales to Kepler*, republished with revisions as *A History of Astronomy from Thales to Kepler*, Dover Publications Inc., New York 1953.
A detailed scholarly history of planetary theory.

HURD, D. L. and KIPLING, J. J. (eds), *The Origins and Growth of Physical Science*, Vol.I, Penguin Books, London 1964.
Contains extracts from the writings of Ptolemy, Copernicus, Kepler, Galileo, and Newton.

KUHN, T. S., *The Copernican Revolution*, Vintage Books, New York 1959.
Discusses the switch from Ptolemaic to Copernican-Newtonian conceptions.

PANNEKOEK, A., *A History of Astronomy*, Interscience Publishers Inc., New York 1961.
A very readable general history of astronomy.

TOULMIN, S. and GOODFIELD, J., *The Fabric of the Heavens*, Penguin Books, London 1963.
Provides accounts of the several planetary theories.

DE SANTILLANA, G., *The Crime of Galileo*, University of Chicago Press, Chicago 1955.
Discusses the scientific and philosophic issues involved in Galileo's clash with the Church.

WOOD, H., *Unveiling the Universe: the aims and achievements of astronomy*, Angus and Robertson, Sydney 1967.
A serious account of astronomy aimed at the beginning student and the thoughtful layman. Chapters 1 to 7 are relevant to the case study presented, though the reader will almost certainly want to read on in order to learn about the other stars, our own galaxy and the galaxies beyond it.

THE PERIODIC TABLE OF ELEMENTS

HALL, A. R., *The Scientific Revolution, 1500-1800*, The Beacon Press, Boston 1954.
A good account of the early history of modern chemistry is given in Chapter II.

HURD, D. L. and KIPLING, J. J. (eds), *The Origins and Growth of Physical Science*, Vol.II, Penguin Books, London 1964.
Contains extracts from Dalton and Mendeleyev.

HOLTON, G. and ROLLER, D. H. D., *Foundations of Modern Physical Science*, Addison-Wesley Publishing Co., Inc., Reading, Mass. 1959.
Chapters 22, 23, 24, 34 and 35 give a clear account in technical terms of moderate sophistication of the historical background to the periodic table, its progressive formulation and the rise of theory now deemed to explain it.

JAFFE, B., *Crucibles: The lives and achievements of the great chemists*, 7th printing, Tudor, New York 1936.
Gives an historical account of Mendeleyev's work and of some of the later theory.

PARTINGTON, J. R., *A Short History of Chemistry*, 3rd rev. edn, Macmillan, London 1960.
Chapters 8, 11, 12 and 15 provide a brief account of the contributions of various chemists from Dalton to Mendeleyev relevant to Case Study 2.

INHERITANCE AND THE GENE THEORY

CARTER, C. O., *Human Genetics*, Penguin Books, London 1962.
Following an initial statement of elementary genetic theory, a detailed account of much of what is known of human inheritance, bodily and behavioural, is presented.

DUNN, L. C., *A Short History of Genetics*, McGraw-Hill, New York 1965.

MOORE, J. A., *Principles of Zoology*, Oxford University Press, New York 1957.
Part B gives a very clear account of the development of genetic concepts from pre-Mendelian days up to about 1950.

PETERS, J. A. (ed.), *Classic Papers in Genetics*, Prentice-Hall Inc., Englewood Cliffs, N.J. 1959.
Contains the greater part of Mendel's formative paper and other basic contributions from Sutton, Morgan, Müller, and Watson and Crick.

SINNOTT, E. W., DUNN, L. C. and DOBZHANSKY, T., *Principles of Genetics*, 5th edn, McGraw-Hill, New York 1958.

SRB, A. M. and OWEN, R. D., *General Genetics*, Freeman & Co., San Francisco 1952.
These two are standard introductory textbooks at university under-graduate level which may be profitably tackled by the serious general reader wishing to learn more about genetics.

STERN, C. and SHERWOOD, E. R., *The Origin of Genetics: A Mendel Source Book*, Freeman & Co., San Francisco and London 1966.
Contains a modern and more accurate translation of Mendel's classic paper.

PHILOSOPHY OF SCIENCE

There is an increasing number of books dealing with the types of problem raised in the preceding chapters. In increasing order of difficulty and comprehensiveness the following are suggested:

TOULMIN, S., *The Philosophy of Science. An Introduction*, Hutchinson's University Library, London 1953.

FRANK, P., *Philosophy of Science. The Link between Science and Philosophy*, Prentice-Hall, Englewood Cliffs, N.J. 1957.

NAGEL, E., *The Structure of Science. Problems in the Logic of Scientific Explanation*, Harcourt, Brace and World Inc., New York 1961.

There are several collections of papers from various sources which will be found valuable:

DANTO, A. and MORGENBESSER, S. (eds), *Philosophy of Science*, World Publishing Co., Cleveland 1960.
Parts I and II are most relevant.

FEIGL, H. and SELLARS, W. (eds), *Readings in Philosophical Analysis*, Appleton-Century-Crofts Inc., New York 1949.
Refer especially to Introduction and Parts II, III, IV, V and VII.

FEIGL, H. and BRODBECK, M. (eds), *Readings in the Philosophy of Science*, Appleton-Century-Crofts Inc., New York 1953.
Refer especially to the Introductory papers and to Parts I, II, III and IV.

Index